WHERE WORLDS COLLIDE

Penny van Oosterzee has a Master of Environmental Studies. She is an ecologist and science communicator based in Darwin. She has chosen to put her academic expertise to the test in private enterprise and runs a multi-disciplinary ecological consultancy firm, EcOz Australia, and an award-winning ecotourism business, Discovery Ecotours. Her previous books include *The Centre*, for which she won the prestigious Eureka Book Prize, and *A Field Guide to Central Australia*.

WHERE WORLDS COLLIDE

THE WALLACE LINE

Penny van Oosterzee

Cornell University Press

Ithaca and London

Copyright © Penny van Oosterzee 1997

All rights reserved. Except for brief quotations in a review, this book, or parts thereof, must not be reproduced in any form without permission in writing from the publisher.
For information address Cornell University Press, Sage House, 512 East State Street, Ithaca, New York 14850.

First published in 1997 by Cornell University Press
Sage House, 512 East State Street,
Ithaca, New York, 14850

ISBN 0-8014-8497-9

Librarian: Library of Congress Cataloguing -in-Publication Data are available.

Printed and bound in Australia by Griffin Press

ACKNOWLEDGEMENTS

I am indebted to the following people. Publisher, Bill Templeman, had faith in the book from the very first and graciously put up with the early delays while I grappled with the immensity of the topic. This book may not have been written if not for his support. Noel Preece, my business partner and partner in life, took on most of the workload from our busy business for several months, freeing my time while I became obsessed with another man, Alfred Russel Wallace. Noel still found time to give advice and constructive criticism on drafts, and provided moral support and encouragement throughout. What can I say about Mary Halbmeyer, my editor, except that she has been terrific—a calming, gentle and professional influence throughout. My son Luke Preece once again gave generously of his time and patience. His unswerving love is an inspiration.

My heartfelt thanks are also extended to Tim Flannery who, despite a hectic schedule, managed to find time to review the work for me. Any factual errors, of course, remain mine alone. I also thank Darryl Kitchener (Western Australian Museum), Colin Groves (Australian National University), Greg Leach (Northern Territory Parks and Wildlife Commission), Ian Metcalfe (University of New England), and Colleen Pyne (North Australian Research Unit) for their generous assistance in providing or gathering information.

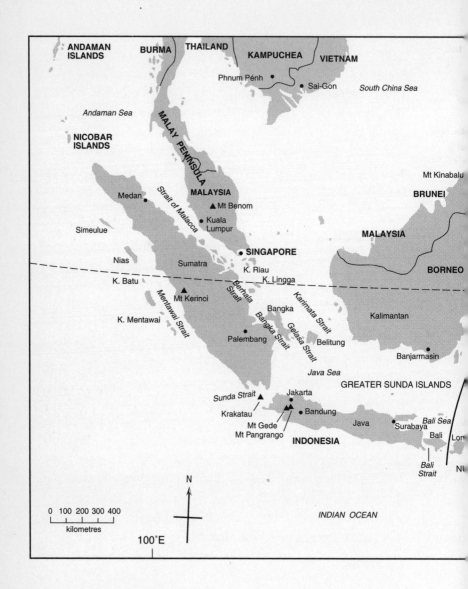

Map 1. The Malay Archipelago.

The islands of New Guinea, Indonesia, the Philippines and the Moluccas.

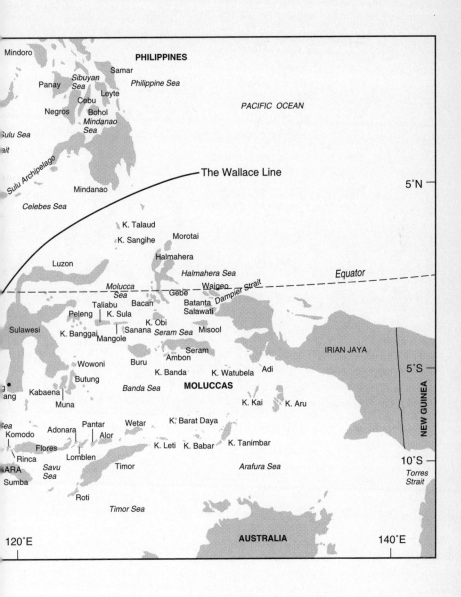

CONTENTS

ACKNOWLEDGEMENTS v
PREFACE xiii

CHAPTER ONE
   BLIND TO THE RISING SUN   1
Noah's Ark—'the species question' and God—Linnaeus and Lamarck—the Sarawak Law and Lyell's *Principles*—Darwin and the Galapagos—Wallace's theory of evolution

CHAPTER TWO
   WALLACE'S LINE   21
Passage to Bali—The megapodes: Orange-footed Megapode, Maleo and Moluccan Megapode—geographic distribution of species—an extraordinary zoogeographic boundary discovered—the Wallace Line—other zoogeographic lines—animals across the Makassar Strait

CHAPTER THREE
   WHERE WORLDS COLLIDE   41
Alice in Wonderland and the formation of the Malay Archipelago—Alfred Wegener and continental drift—plate tectonics and continental reconstructions—rocks as a window into ancient environments—Asia in Gondwana—the global jigsaw puzzle and evidence for the Wallace Line—'looking glass' cicadas

CONTENTS

CHAPTER FOUR
FIRE-SPITTING MOUNTAINS   69

Volcanic islands of Indonesia—the Spice Islands—Wallace and volcanos: anything that did happen, can happen—plate tectonics and volcanos—Krakatau

CHAPTER FIVE
STEGOLAND   83

Present-day ice age—pollen as a clue to glacial impact: Lynch's crater—Pleistocene mammals of Indonesia and the Wallace Line—swimming pygmy elephants and dwarf stegodons—plant boundaries—influence of climate on distribution

CHAPTER SIX
ISLANDS IN THE SEA   97

Avifauna and the boundary between Bali and Lombok—island biogeography—the molecular bird family tree—poor dispersal ability of mammals—the endemism of islands—Moluccan productions—biological mimicry: the oriole and the friarbird

CHAPTER SEVEN
ISLANDS IN THE SKY   123

Carnivorous plants—diversity of tropical mountain environments—ecological 'waifs'—migration pathways of temperate plants—endemism of tropical-alpine species—Mount Kinabalu—climate change and rapid evolution

CHAPTER EIGHT
   THE ULTIMATE ISLAND   135
Sulawesi: individual and contradictory—evidence for a
permanent separation—endemic mammals of Madagascar—
the African connection—the endemics: Babirusa, anoas,
primitive cuscuses, the Sulawesi Palm Civit, tarsiers—
endemism of the mammals: macaques, squirrels, shrews, murid
rodents—a dumping ground for primitives—study of island
dynamics: the Philippines—antiquity of Sulawesi

CHAPTER NINE
   THE BUTTERFLY EFFECT   159
Chaos of the natural world—butterflies in Indonesia—
butterfly biology—the development and evolution of butterfly
wing patterns—Mendel and genetic science—mimicry in
butterflies and the 'supergene'

CHAPTER TEN
   THE RED APE OF ASIA   177
Wallace and Orang-utans in Borneo—mammal curiosities of
the Sunda Shelf: lemurs, tupaias, pangolins—oceanic versus
continental island fauna: every niche filled—carnivores: the
tiger—behaviour and natural history of Orang-utans—
evolution of Orang-utans—plate tectonics and Orang-utans—
Orang-utans and humans

CONTENTS

CHAPTER ELEVEN
WALLACE IN WONDERLAND   197
The Great Pied Hornbill—Wallace's intrepid boat journeys—journey to Seram—richness of Aru—birds of paradise

EPILOGUE
'THEY HAVE ALL LATELY GONE'   217

REFERENCES   221
INDEX   229

PREFACE

THIS IS a book about a great man named Alfred Russel Wallace, the father of biogeography. It is about the field of biogeography and the development of related fields, such as evolution, genetics and plate tectonics. The thread welding these topics together is the discovery of the Wallace Line, a faunal barrier separating the Oriental from the Aus-tralian: monkeys from kangaroos, and pheasants from parrots. This invisible barrier is today as controversial as it was one hundred years ago when Wallace used it to galvanise his Theory of Evolution which forced Charles Darwin's hand.

My account is based on Wallace's own great natural history travelogue, *The Malay Archipelago*, a book he wrote after spending the period between 1854 and 1862 in Malaysia, Indonesia and New Guinea, often in areas where Europeans had never been. Wallace's lively book focuses this story. As much as possible I have discussed those concepts and natural histories with which Wallace himself grappled. Wallace, for instance, was not so much interested in flora—except for the flora of tropical mountains—and, accordingly, where flora is mentioned in this book, it is to provide continuity and completeness. The seemingly esoteric topic of butterfly wing patterns, on the other hand, receives a chapter because Wallace was so besotted by them.

This comparative approach has allowed for some fascinating contrasts between nineteenth-century and twentieth-century ideas and scientific developments, such as the fact that chaos and evolution is, in fact, written in the patterns of butterfly wings.

To create a lively narrative, which surely reflects the reality, I have taken some licence and created scenes—such as Wallace

PREFACE

catching insects while deeply pondering evolution—which are drawn from my imagination.

This book has been written in order to maximise readability, and therefore referencing has been kept to a minimum. All Wallace's quotes and thoughts, too, unless otherwise cited, are drawn from *The Malay Archipelago*. Some of the references that are cited I have drawn on considerably, and I commend them to anyone for further reading.

CHAPTER 1

BLIND TO THE RISING SUN

C'mon let's get back to Creation! I can't waste any more time! Otherwise they'll think I've lost control again and put it all down to evolution.

The Supreme Being. Monty Python's Time Bandits, Handmade Films, 1980.

ALFRED RUSSEL WALLACE was up against it. His gun lay propped against a fallen, rotting tree. A stoppered killing-bottle with alcohol stood next to it. He looked about for any movement—a butterfly, a bird—in the bright and varied foliage of the virgin forest growing at the limits of the clearing. Wiping the sweat from his brow he carefully folded his butterfly net, laying it neatly on the ground and stooped once again to sift his large, gentle hands carefully through the rich litter of timber, bark and trees, which had accumulated in the steamy, sunlit clearing. Wallace probably didn't notice the sweat dripping from his eyelashes onto his glasses as his focused, short-sighted, gaze beheld yet another beautifully patterned beetle; a new species for sure. Never before or since had he witnessed such a flagrant, seemingly purposeless variety of life. His lively blue eyes sparkled at the possibilities such a phenomenon suggested. It left him breathless. It was also against The Law. And he liked that.

From 20 April 1854 to 1 April 1862, Wallace trekked the Malay Archipelago, today's Indonesia, wallowing in its wildlife. Situated on the equator and bathed by the tepid water of three great tropical oceans, this area of 13 000 islands displayed an unheard-of, wonderful variety of species. Wallace had first become fascinated by beetles when a young man in England, after seeing the collection of his friend Henry Walter Bates (1825–1892), who became a famous biologist in his own right after developing the concept of mimicry. As a youth, Bates had collected and carefully pinned hundreds of beetles from the English countryside. Here, in the Malay Archipelago, beetles existed in an almost infinite number of specific forms, leaving Wallace dazzled.

Such variety was by no means restricted to insects. Indeed it was the promise of such multiple wonders in unknown lands that drew Wallace to the Malay Archipelago. As a young 'fly-catcher' (as one of his peers ungraciously called the specimen collector), Wallace had earlier completed an adventurous collecting trip to the Amazon from 1848 to 1852, being the first European to penetrate the upper reaches of the Rio Negro. He emerged from the jungle near death with dysentery and malaria, but with an incredible collection of Amazonian animals. On the return trip to England the ship which carried him caught fire and sank, taking with it his irreplaceable specimens and very nearly his life.

The awkwardly shy Wallace was not shy of adventure. However, his total inability to participate in sophistry in a very sophisticated Victorian England; his intense interest in the origins of species; plus the inescapable fact that he was broke as a result of the shipwreck, determined that his next destination should be nothing less than the most remote, least collected region of the world. Wallace had learned that the first visit by a naturalist to the Malay Archipelago had occurred only in 1776 and there had been virtually no exploration since. It was left to Wallace in 1854, at the age of 31, to open up this new world to science.

The region was steeped in mystique. The richest variety of fruits and the most precious spices were indigenous there. Wallace had heard of the monstrous flowers called *Rafflesia* that are 'three feet wide and weigh 24 pounds!'. The great green-winged Ornithoptera, which Wallace later dubbed 'the prince among butterfly tribes', the man-like Orang-utan whose origin sparked Wallace's curiosity as had no mammal other than man himself, and the bird of paradise which had never been exhibited in Europe, were other features of the archipelago's giddy, divergent forms of life. Indeed, it was this seemingly endless variety of life that gave Wallace's life meaning.

The endless modification of a species' structure, shape and colour, particularly in insects and their innumerable adaptations to diverse environments were, in the early nineteenth century, inexplicable. At this time, species were considered fixed and unvarying. It was a problem which had taunted Wallace since his explorations in the Amazon. These chaotic outbursts of life did not fit at all comfortably with the Christian conception of the world which demanded a Divine Plan with underlying Law and Order. In 1854, as Wallace stood in a tropical sunlit clearing, gazing in wonderment at one new species after another, he was pondering thoughts that were not at all allowed for under the contemporary Christian doctrine.

In the early and mid-nineteenth century, virtually all scientists accepted the traditional Christian concept of a Creator God. The Bible was regarded as a literal tome, a real account of the origins of the natural world. Myth was fact. So was the view of Aristotle, who had held that different types of animals were arranged up a ladder, with Humanity, being the pinnacle of divine success, on the top rung. Only by the Hand Of God could this static situation change and an animal be spirited up the ladder.

Many still believed that the geographical distribution of animals was explained by Noah's Ark and its final abandonment on Mount Ararat, from whence every species of today's animals trudged, flew or swam to their current position on the face of the globe. In the sixteenth century, only 150 kinds of mammals, 30 snakes and 150 bird types were described. There was room for all in the Ark. Unfortunately, however, foreign travellers began to bring back reports of an unprecedented wealth of animals in the New World of America.

These discoveries proved a touch vexatious, since the startling differences between animals inhabiting America compared with Europe were biblically inexplicable. In the mid-sixteenth century, Edward VI, a close confidant of God, was forced to clarify the situation explaining the unquestionable fact that 'God of heaven and earth greatly providing for mankind, would not that all things should be found in one region, to the end that one should have need of another'.[25] Nevertheless, questions remained. How did animals manage to get to the New World from Mount Ararat and why did none stop to settle in between? Why were there so many strange creatures in America and how did they all fit in the Ark? Most importantly, why would God have created the beasts of America since they were by and large beasts of prey and noxious animals, of no use at all to man for whom all creatures were made? There was not even that necessary creature, a horse.

By the end of the seventeenth century there were over 500 species of birds, 150 quadrupeds and 10 000 insects categorised. The Ark was getting crowded. Naturalists of the day largely refused to confront the dilemma, saying that it was only man's limited imagination that prevented him from seeing how it could be done.

One natural philosopher, Isaac de La Peyrere, had trouble with all this. In 1655 he suggested the radical idea that perhaps the Flood wasn't universal. Perhaps it was a local event confined to Europe and the Middle East—the Biblelands. Other races of people, also deemed to have started with Noah's sons (the African Negro was descended from Ham, for instance, the one son who fell out of favour and was destined to die cursed), and the birds and beasts and plants could have lived on, untroubled by Catastrophe. He suggested that Adam might not have been the first-created man, but merely the first of the Jewish race. China, America, Greenland and the mysterious continent to the south might not have been involved with the rest of the biblical story. To La Peyrere's mind, this at least better explained the evidence of the natural world.

This was too much for seventeenth century Europe. La Peyrere had gone too far! He was arrested in 1656 and escorted to Rome to sign a public retraction in the presence of Pope Alexander VII.

Even though La Peyrere had been punished and men still continued to seek Absolute Design in the universe, La Peyrere had, nevertheless, placed a crack in the legend of the Ark. At the dawn of the Age of Exploration, as the natural history cabinets of Europe were filling to brimming with specimens of new animals and plants, the concept of a literal Ark began to sink.

The great naturalist philosopher, Carolus Linnaeus (1707–

1778), was born into this era of enlightenment. A product of his time, for Linnaeus the natural world was the pathway to understanding God's work. Linnaeus felt himself chosen to describe and catalogue the natural world, in order to fully understand His Laws and to seek the Divine patterns of order and regularity. Linnaeus invented the binomial system of classification of animals and plants, using Latin names—the system we still use today.

Linnaeus' task was monumental. He had to come to grips with the avalanche of new animal and plant types, what he called 'species', including the 5600 that he personally named and catalogued. It was obvious to him that no literal Ark could have carried such a number and Linnaeus did away with the idea. Linnaeus reasoned that Mount Ararat itself must have been the metaphorical Ark and that all creatures were created on it. They weren't, however, just a chaotic collection thrown helter-skelter on a mountain. Instead, he argued, the mountain had a wide range of ecological belts, from the polar to the tropical arranged around it. This allowed for each pair of animals to be created in the climate most suitable for them. Creative powers also determined the number of each type of animal and plant and also the numerical relations of flowers and animals in any one region. The form of an animal was also preordained. Birds' beaks, for instance, had been constructed especially for the type of food on which they fed and no other explanation was either possible or needful.

From this single, original source mountain, the animals moved to latitudes and environments where they would remain. For Linnaeus the result was pleasing. Each animal of the present day was descended from the original pair. All animals and plants were perfectly constructed for the environment in which they were situated. Each plant and animal had a station and a function. The whole system of plants and animals and their interactions were held in a delicate balance, driven by Divine Laws; the system throbbed with a predestined vibrancy.

Linnaeus emphasised the fact that God had created the laws which drove these complex interrelated systems and the species within them. Perhaps inadvertently this resulted in the attitude that since God had created the laws, it was okay to study them. As well as giving tacit support to study ecosystems, Linnaeus' approach was important for modern biogeography because it highlighted the importance of the adaptation and suitability of individuals to their environment.

Inevitably, Linnaeus' interpretation opened the door for more secular accounts of geographical distribution. These accounts ridiculed the idea that only one pair of each kind of animal was created. Surely, for instance, the Lion would quickly eat all the herbivores in Linnaeus' living museum and soon die once it had eaten everything in sight. Surely it was more rational to believe, as La Peyrere had done, that every animal was created in the area in which it now lives, under the same climate that it now enjoys. This idea—that entirely and excitingly different suites of species were created in different regions of the Earth—fired the imagination of a whole retinue of 'apostles' who journeyed far and wide to collect samples of species from the furthest reaches of the world, compiling regional lists of flora and fauna.

It was soon confirmed that plants and animals did indeed fall into natural associations. Botanist Alexander von Humboldt (1769–1859) reasoned that a glimpse into the Divine Plan could be had by the comparison of one botanical region with another. In the early part of the nineteenth century, this comparison was carried out by merely listing the plants in region A and all those in region B, leaving it to the reader to extract whatever information was required. No-one asked why, or what statistical methods were used, or what the figures actually meant for natural history. The object was to gather facts, not fool around with theories. In the Malay Archipelago, Wallace bore the brunt of this attitude in a letter from his agent, Samuel Stevens, who told Wallace (who published extensively while away) to stop theorising and just get on with collecting the 'facts'.

Meanwhile the number of new species being discovered increased at a staggering rate. As many as 161 000 species that were unknown in 1750, could be found in collections in the year 1833.[10]

In 1800, the French naturalist Jean-Baptiste de Lamarck (1744–1829) developed a novel way of splitting this burgeoning animal kingdom: into vertebrates and invertebrates. He suggested that this split reflected the relationship between the two: that invertebrates *evolved* into vertebrates. It was a startling idea.

'I do not doubt that ... water is the true cradle of the entire animal kingdom.

We still see that the least perfect animals and they are the most numerous, live only in water ... that it is exclusively in water or very moist places that nature achieved and still achieves in favourable conditions those direct or spontaneous generations which bring into existence the most simple organized animalcules, whence all other animals have sprung in turn.

... After a long succession of generations these individuals, originally belonging to one species, become at length transformed into a new species distinct from the first.'[40]

Other naturalists had also questioned the divinity of nature but they were by far in the minority and were fearful of the formidable, righteous opposition. Erasmus Darwin (1731–1802), Charles Darwin's grandfather, for instance, also believed in the concept of evolution, but disguised the heretical thought in a frivolous poem.

> Organic life beneath the shoreless waves
> Was born and nurs'd in ocean's pearly caves;
> First forms minute, unseen by spheric glass
> Move on the mud, or pierce the watery mass;
> These, as successive generations bloom,
> New powers acquire and larger limbs assume;
> Whence countless groups of vegetation spring
> And breathing realms of fin and feet and wing.
>
> *Erasmus Darwin. 'The Temple of Nature', 1803.*

It was clear to Lamarck that he lived in a dynamic world, where the climate, geology and geography were constantly changing, sometimes drastically. If there had been a single godly creation, then all the organisms would be poorly adapted indeed to their environment. Since animals were actually very well adapted to their environment, Lamarck reasoned, then they must have a capacity for change. They must evolve.

But it was here that Lamarck lost the plot. Being a devout man, he believed that the Sublime Author created the Laws for evolution to occur. Lamarck believed that animals evolved toward 'perfection', where humanity represents the highest perfection that nature could attain. Even though Lamarck spoke about branches among evolutionary lines, he still managed, somehow, to believe in the fixity of species up the Aristotlean ladder. In clinging to the biblical apron-strings, Lamarck had to perform some incredible theoretical contortions, such as inventing the concept of spontaneous generation to spirit into existence simple 'infusorians' to replenish the bottom rung of the ladder as animals and plants moved up the rungs.

Evolution, Lamarck said, occurred through a process of 'transformation'. Take, for instance, a bird living close to water and not wanting to swim. It would constantly stretch its legs so that over time it would evolve into a long-legged stork. Lamarck believed that this sort of transformation came about by the internal movement of fluids. The more complicated the animal and the more 'willing' the animal was to change, the more rapidly did these 'internal fluids' flow and the more could the animal change. Somehow these changes would then be inherited by the animal's offspring. Lamarck believed in a Divine Catalyst to account for the origin of species.

The idea, of course, had germs of the truth, but mid-nineteenth century naturalists would have none of it, scoffing at the concept of species changing into other species. With no evidence of how such a bizarre thing could come about and with 2000 years of belief in an Ordered World, regulated by a Superior Wisdom, behaving in a set, one-track way toward excellence (as personified by Man), they resorted to Creation, which, of course, required no evidence. One of the great scientists of the mid-nineteenth century, Louis Agassiz (1807–1873), described the obligations of naturalists as complete as soon as they have proved His existence. In 1857, he said that 'natural history must, in good time, become the analysis of the thoughts of the Creator of the universe, as manifested in the animal and vegetable kingdom'.[1]

Leading naturalists railed against the concept of evolution. They held that everything in nature was planned, designed and had a predetermined end. Everything in nature revealed glimpses of the Supreme Intellect, the Reflective Mind. Nothing was impossible in creation. When a species became extinct it was simply replaced by a new species. There was no essential variation between individuals of a species and hence species did not evolve.

This was the era into which Alfred Russel Wallace stepped. Wallace, an impious innovator, with a free-ranging, undisciplined and unstructured mind, was not satisfied with this and other vague solutions to the vexing problems of the origin of species. Instead of a static and neatly ordered world where animals formed distinct and clear species, always fixed in form, he noticed outbursts of variation within species. Clearly there was another principle regulating the infinite variety of life. Like his contemporary Charles Darwin (1809–1882), Wallace was ever turning over 'the species question'.

Men like Agassiz were victims of the thoroughness of their education. Alfred Russel Wallace, on the other hand, was born (in 1823) into a family of no distinction, in an era when individuals were locked into their family's social status, as fixed and permanent as the conventional ideas of species. Wallace went to school till the age of 14, when he was forced to leave because of his father's bankruptcy. Ironically for the future founder of biogeography and expert at Latin designations, his two most painful subjects were Latin and geography. Nevertheless, Wallace was an avid reader who devoured books on a multitude of topics and enjoyed way-out ideas, such as socialism and evolution. When he left for the Malay Archipelago he was essentially a self-educated drifter without prospects of regular employment.

Wallace scoffed at the belief that animals had been created entirely for the benefit of Man who was, alone, graced with an appreciative and aesthetic sense. This view held that the elephant, for instance, had been created docile so that humans could use it. And it had been created in exactly such a region where it could find suitable food for itself.

From the example of English trees and fruits, it was plainly obvious that small fruits always grew on lofty trees, so that their fall would be harmless to humans. Large fruits trailed on the ground. In the Malay Archipelago, Wallace pointed out that two of the largest and heaviest fruits known, the Brazil-nut and the Durian, grow on lofty forest trees, from which they fall as soon as they are ripe, wounding or killing any hapless native inhabitant wandering beneath. 'From this we may learn two things' he said dryly, 'first, not to draw general conclusions from a very partial view of nature and secondly, that trees and fruits, no less than the varied productions of the animal kingdom, do not appear to be organized with exclusive reference to the use and convenience of man.'

The constant discovery of new species, beginning in the Amazon and continuing, in a biological eruption, in the Malay Archipelago, altered Wallace's perceptions of humanity and its place in the universe. He thought the unthinkable: Man was not at the centre of world, any more than the Earth was at the centre of the universe.

Wallace knew well the implications of questioning the origin of species. He knew that he was grappling with the whole question of the organic world and its connection with a time past and with Man. He was grappling with basic beliefs, with religion and with ethics.

Einstein's theory of relativity could hardly affect anyone's personal belief. The Copernican revolution—that the Earth revolves around the sun, rather than the other way around—and Newton's world view of motion and gravity, required some readjustment of traditional beliefs. The mystery of the origin of species required a rejection of basic belief systems; in a sense the murder of God.

In 1844, in a letter to the famous botanist Joseph Hooker, Charles Darwin had also written as much. 'I am convinced ... that species are not (it is like confessing murder) immutable.'[14] Not that Wallace, at the time, knew this of Darwin, who, though already a well-respected, well-published and socially well-placed naturalist, was 15 years away from publishing any of his ideas regarding the origin of species.

Likewise, Charles Darwin had barely heard of the young Wallace, being aware only vaguely of his adventures in the Amazon. Both men, however, were deeply fascinated in the peculiarities and irregularities in animal distribution. Both realised that living things were elaborately adapted to their environment, in form, colour and habit. Both shunned the absurdity of creationist views, which would have new species sprinkled, wand-like, in areas becoming depleted. Both believed in evolution.

Unlike Darwin, however, Wallace was not at all daunted by the possibility of ridicule by his peers; equally, he himself was ready to ridicule the often ludicrous metaphysical ideas of his time. A paper written by Professor Edward Forbes (1815–1854) in 1854, suggesting that the wealth of life on earth oscillated, for no particularly good reason, from rich to poor through geological eras and that any resemblance of groups from one era to the next was due to a penchant of God for one design over another, was like a red rag to a bull for Wallace. Annoyed with these convoluted and absurd ideas when a simple, secular hypothesis could explain all the facts, Wallace sat down in Sarawak to write a paper titled 'On the Law Which Has Regulated the Introduction of New Species'; a law which came to be known as the Sarawak Law.

The essence of the Sarawak Law is that 'every species has come into existence coincident both in space and time with a pre-existing closely allied species'. It was as close to an explanation of evolution as anyone had yet dared.

Contrary to Forbes, who thought that each geological period was marked by the introduction of totally different forms from those which came before or after, Wallace pointed out that the distribution of animals and plants in space echoes their distribution in time. Higher taxa—classes and orders—are generally spread over the whole globe. Primates, for instance, are found everywhere except Australia and Antarctica. This is in effect an echo of their past distributions. Genera and species, more often than not, are specific to an area, often a very small area. In particular, closely allied species will be found geographically near to each other. For example, Lemurs are only found in Madagascar and Orang-utans only in Indonesia.

Wallace's one great advantage was that he knew what he was

talking about. Daily, new species of butterflies were swept into his net; hourly, new beetles crept over his fingers; he watched dazzling new birds create rainbows in the trees. With this constant bombardment of the 'facts', Wallace recognised that species were the fundamental units of evolution, being both the products of speciation and the things which are thought to speciate. He saw that species were so numerous and their modifications of form and structure so varied, that the tree of life could not possibly be a ladder. Rather the lines of relationships were, he wrote, 'as intricate as the twigs of a gnarled oak or the vascular system of the human body'.

Wallace's friend Bates, still in the Amazon after eight years, was also at the front-line of the species question. When he read of the Sarawak Law he penned an enthusiastic note to Wallace. 'The idea is like truth itself, so simple and obvious that those who read and understand it will be struck by its simplicity; yet it is perfectly original.'[7]

In fact, the Sarawak Law was the closest anyone had come to providing at least part of the answer to the origin of species. It suggested the when and the where of evolution: that evolution, from geographically nearby and closely related species, was occurring constantly. But the greatest mystery of the species question, the how, was still not answered.

The Sarawak Law was based on the work of Sir Charles Lyell (1797–1875), one of the most important and famous scientists of the era. From 1830 to 1833, in his *Principles of Geology*, Lyell assembled overwhelming evidence in favour of inorganic evolution: that the present state of the inorganic world was the result of a constant series of changes that had been going on since the earliest periods and were still going on. Lyell's view depended on a strong continuity between the present and the past. He felt that the earth had a history of never-ending fluxes, where each epoch supported a great number of distinct populations that expanded and contracted their boundaries as geographical barriers were removed or erected.

Lyell took the important step of assuming that modern floras and faunas had all enjoyed lengthy histories. This sort of thinking was an innovation in an era still very much constrained by thoughts of an Interventionist Deity. Yet, in what was truly extraordinary and in the face of overwhelming evidence against it, Lyell managed to cling tenaciously to the belief that species were still created at appointed times and hence were fixed and unchanging, with no variation. He somehow imagined countless small catastrophes that wiped out species so that a very hands-on Creator would have to be constantly restocking His world over time with new species, which, moreover, looked similar to the ones just vanished. In other words, Lyell did not extend his ideas of inorganic evolution to the organic world. Both Alfred Russel Wallace and Charles Darwin did.

Darwin's inspiration also had its origin in Lyell's Principles, which Darwin thought was a masterpiece of interconnected reasoning. Lyell's Earth was forever on the move. Unlike Lyell, however, Darwin realised how markedly the living world was affected by geological processes and how delicately balanced were the relations between the organic and inorganic kingdoms. Darwin realised that biological arguments were crucial to the overall strength of a truly dynamic, historical explanation of the Earth.

Nevertheless, Darwin, when he began his now-famous journeys in the *Beagle* to South America in 1831, was a creationist, like every other naturalist. He had no reason to question this most fundamental of contemporary philosophies. Until he saw the Galapagos Islands.

In September 1835, Darwin was looking forward with building excitement to examining the Galapagos. Here was a constellation of new volcanic islands ready to experience all that the world could offer; here was a new scientific experience.

Darwin was going to explore how new lands were peopled with living beings. Was it by immigration or by creation? Did new lands have new species? Contemporary theories of the day allowed Darwin two choices. Either the Galapagos was going to be peopled with displaced members from South or Central America, or, more likely, being oceanic islands far away from the mainland, they would sport their own independent creations.

Perplexingly, he discovered that the Galapagean species were neither one thing nor the other. They were the same and yet not the same. Faced with the existence of insular tortoises, mockingthrushes and finches, like, but not like, those of South America, Darwin began to question the most fundamental of contemporary theories. By the time he returned to England from the Galapagos in 1836, Darwin had been reborn: it was clear to him that species change and evolve, giving rise to other species.

Darwin's metamorphosis from creationist to evolutionist was complete. From then and for the rest of his life 'the species question' dominated him. He searched everywhere in the literature for a mechanism for evolution, hitting upon the idea of natural selection in 1838 after reading an essay not about wildlife but about humans. The 1798 essay by Thomas Robert Malthus (1766–1834) explained that it was natural checks that kept human populations in control. Things like war, famine and natural disease culled human populations.

At last, a theory to work with! Natural selection, the survival of the fittest. As with humans so it must be for the natural world. Combined with some sort of isolating mechanism, such as geological change, like the formation of islands or mountains, natural selection would result in new species, descending from a common ancestor.

But this was the beginning, not the end, of the species question for Darwin. He spent the next 20 years researching, reflecting and developing his theory of evolution. Darwin always carried around with him red notebooks which he filled, one after the other, with observations to support his new ideas. At home he experimented extensively with animals and plants. He had tanks in the cellar in which he floated seeds to see if they could withstand salt water and so show that they could disperse to new lands via sea currents. He floated a pair of duck's feet in a tank with freshwater snails to see whether they would climb aboard, which they obligingly did, and so be transported to new geographical areas, shaking the feet vigorously to try to dislodge the snails, no doubt to the amusement of his children. In particular he spent eight years examining living and fossil barnacles in the minutest possible way, to show that variation was a natural consequence of reproduction. He examined 12 volumes of botan-ical data from around the world and tediously applied what he called botanical mathematics to groups of genera to prove his 'principle of divergence' (the long sought-after crowning glory of his origin of species), that species evolved, constantly branching away from a common ancestor into new species, new genera and finally, over time, into new families. Darwin never left any question unanswered and his writings were meticulous and often laboured in order to cover every possible slant. Today we might say that he was bogged down in the problem.

In 1856, galvanised by the publication of the intriguing Sarawak Law by Wallace, Darwin's friend Charles Lyell urged him to publish without delay, pointing out the similarity of the two approaches. Darwin took the advice and started writing the 'big species book'.

A paragraph in a letter to a colleague in America, Asa Gray (1810–1888), in September 1857, is one of the most succinct and closely argued explanations of his findings:

'The varying offspring of each species will try (only a few will succeed) to seize on as many and as diverse places in the economy of nature as possible. Each new variety or species when formed will generally take the place of and so exterminate its less well-fitted parent. This I believe to be the original of the classification or arrangement of all organic beings at all times. These always seem to branch and sub-branch like a tree from a common trunk; the flourishing twigs destroying the less vigorous; the dead and lost branches rudely representing extinct genera and families.'[10]

In June 1858, when his work was nearly complete, Darwin was rocked to his very foundations. An unexpected letter from Wallace, who was in Ternate in the Malay Archipelago, was delivered to Darwin. It contained a paper titled 'On the Tendency of Varieties to Depart Indefinitely From the Original Type' and it outlined in a clear and succinct manner the origin of species by natural selection, explaining the tendency of species to diverge from a common ancestor.

What had taken Darwin virtually all his life, Wallace had done during a bout of feverish malaria. '... in a sudden flash of insight, it was thought out in a few hours, was written down with such a sketch of its various applications and developments as occurred to me at the moment, then copied on thin letter paper and sent off to Darwin—all within a week.'[43]

It was a coincidence that still grips the imagination. We can only imagine how Darwin felt as he fingered the flimsy paper holding the ideas, almost to the word, that were his life-blood. Wallace had written to Darwin as the only other person he knew who was interested in the species question.

In a diplomatic coup, the joint 'Theory of Evolution' by Charles Darwin and Alfred Russel Wallace was announced in the Linnean Society meeting in London of July 1858. Why did some members of a species die while others lived? Bates and others later kicked themselves in public in astonishment and embarrassment over the apparent simplicity of the solution. Since recorded history the world's greatest thinkers had been blind to the rising sun.

For a while Darwin must have forgotten the heretical implications of the Theory. The reception of the July meeting was ominous in its silence. Wallace and Darwin wrote for a generation which poured contempt on those who upheld the derivation of species from species by any natural law. In November 1859, Darwin, coerced by Wallace and Lyell, finally published the *Origin of Species*.

Then all hell broke loose. In Rome, Pope Pius IX placed the book on the *Index Expurgatorius*. In Britain, Cardinal Manning organised a society 'to fight this new, so-called science that declares there is no God and that Adam was an ape'. The Jews were rattled. 'Darwin's volume is plausible to the unthinking person,' Rabbi Hirschberg, an American, sputtered, 'but a deeper insight shows a Mephitic desire to overthrow the Mosaic Books and bury Judaism under a mass of fanciful rubbish.' Protestants were equally alarmed. The Bishop of Oxford, writing in *Quarterly Review*, charged that the concept of interminable natural selection 'attempts to limit the power of God ... the bible and dishonours nature'. Free-thinkers of no fixed religion were equally up in arms. 'The one motive of the whole book is to dethrone God.' Was Darwin proposing that Queen Victoria was related to an ape?[9]

In the Malay Archipelago, Wallace, unaware of the fuss, was chuffed. He had been welcomed into the select fraternity of naturalists whose interests took them beyond the mere description of species.

CHAPTER 2

WALLACE'S LINE

I believe the western part to be a separated portion of continental Asia, the eastern the fragmentary prolongation of a former Pacific continent.

Alfred Russel Wallace in a letter to Henry Bates, 1858.

IT WAS late May, 1856. Wallace walked the narrow, colourful streets of Singapore, his eyes alive with the sights only the races of the exotic East could provide. Plump, round-faced Chinese merchants strolled by importantly, their long tails of hair, neatly plaited and tied with red silk, swaying as low as their ankles. 'Klings', from western India and Arabs, the merchants and shop-keepers of the time, eyed prospective passers-by, inviting them in to inspect the merchandise and asking twice the amount they were really prepared to take. Cotton, penknives, corkscrews, gun-powder, writing-paper—any article could be purchased cheaply, as Wallace knew. Javanese sailors and traders from Sulawesi and Bali bargained carefully, the mock-destitute expressions on the shop-keepers' faces good signs of success.

Tailors and shoe-makers worked busily at their tables. Barbers shaved heads and cleaned ears, the latter operation requiring a great array of tweezers, picks and brushes. All about the streets were sellers of water, vegetables, fruit and soup, their cries unintelligible in any language. Wallace smiled good naturedly, a friendly nod here, a word of greeting there, bowing and refusing politely to walk in and share tea with merchants he had previously bought goods from. He was on his way to the harbour.

As usual, Singapore harbour was crowded with men-of-war and trading vessels of many European nations. Hundreds of Malay praus and Chinese junks, from vessels of several hundred tonnes, down to little fishing boats and passenger sampans, dotted the harbour, their sails like so many brown and white-winged butterflies, flags like colourful eyespots. Wallace was looking for a vessel going to Makassar, today's Ujung Pandang, situated on the south-eastern arm of the spider-shaped island of Sulawesi. He had spent two years in Borneo, Malaysia and Singapore and was eager to explore further. Several times over the past several months he had come down from the forest-clad, tiger-inhabited hills in the centre of Singapore island, seeking unsuccessfully for a direct passage to Makassar. And today was no different. There were no ships going directly to Makassar. There was, however, a Chinese schooner going to Bali, a Hindu island off the east coast of Java. An island he had never intended visiting. From Bali he could find his way to Lombok and then, eventually, to Makassar.

How different history would have been if Wallace had not gone to Bali. The sharp faunal break he was to discover between

Bali and Lombok would not have force-grown his embryonic ideas about the geographic distribution of animals and about evolution. Had he not gone to Bali, he may not have written to Charles Darwin with his Sarawak Law and then with his theory of evolution, which finally forced Darwin's hand. Had he not gone to Bali, the world would have been forever deprived of the romance and controversy of the Wallace Line.

The *Rose of Japan* belonged to a Chinese merchant, was crewed by Javanese sailors and was commanded by an English captain. Wallace took a passage on her on 25 May, arriving on the north side of the island of Bali 20 days later, on 13 June, 1856.

The island was stunning. The whole surface of the land was divided into irregular patches, terraced to follow the contours of the landscape and elaborately irrigated with ditches and channels, which diverted the water that descended from the mountains. Wallace had never beheld so beautiful and well-cultivated a district outside Europe. The sophistication of the island was bewitching and its landscaped facade was a veil blurring the importance of its natural history. Spellbound, Wallace neglected collecting some specimens of animals that he would never meet with again. One was a bird, the Asian Golden Weaver *Ploceus hypoxanthus*, a golden-yellow finch with a black face, which builds bottle-shaped nests. Like so many birds of Bali, it is an Oriental bird, here at the extreme easterly limits of its range.

After a few days, Wallace sailed to Lombok from where he would await a boat going to Makassar. The strait between Bali and Lombok is only 25 kilometres (15 miles) wide, slicing between two volcanos, perfect cones rising out of the mists. From the water, Wallace watched them glowing with the rays of the evening sun at the end of a tropical day. It must have been a magical moment.

After a while, the giant swells of the Great Southern Ocean, squeezing through the deep strait, dumped a relieved Wallace onto the steep beach of Ampanam Bay. Walking up the beach, away from the thunderous sound of the rollers which toppled onto the black volcanic sands, Wallace heard a strange, loud bird call. The locals called the bird, onomatopoetically, 'quaich-quaich'. It was a Helmeted Friarbird *Philemon buceroides*, closely allied with the friarbirds of Australia.

Where were the Oriental barbets, fruit-thrushes and woodpeckers, the same birds he had seen in Malaysia, Borneo and Bali, the latter still clearly visible across the strait? Instead the forests of Lombok echoed with the loud strangled screams of Australian cockatoos, and honeyeaters flitted through the trees. For the first time he also met with the strange mound-making bird *Megapodius reinwardt*, a megapode (literally 'big foot') which industriously uses its big feet to rake together rubbish also found in Australia—dead leaves, sticks, stones, earth, rotten wood—till it forms a large mound, perhaps two metres high and four metres wide. In the middle of this it lays its brick-red eggs, the rotting material of the mound producing heat, like an oven, to incubate the eggs.

Common enough among reptiles, in birds the use of environmental heat to incubate eggs is known only among the megapodes. Not that this means that mound-building is a relic of an ancient reptilian past. On the contrary, it seems clear that the megapodes have descended from a fowl-like ancestor which used body heat to incubate eggs on the ground. This Gondwanan jungle-nester probably scratched litter over the eggs whenever it left the nest. Gradually, more and more matter was heaped on the eggs and the adults returned less and less. Occasionally, eggs would hatch in the absence of adults and natural selection would favour the most precocial chicks.

To most nineteenth-century, European natural historians, birds were preordained to be parentally affectionate. The whole idea of them abandoning their eggs in a cold and calculating and, hence, unbirdlike manner, was unacceptable. Therefore strange stories trickling back from the Far East of fowl-like birds laying their eggs in volcanos, or burying their chicks in mounds, were dismissed as myths. Naturalists, however, aided by local people, soon discovered that there were indeed birds that laid their eggs in volcanically heated sands, or buried their eggs in mounds containing many tonnes of material which the birds built themselves. In fact, the paramount evolutionary achievement of the megapodes has been their utilisation of naturally occurring heat for the incubation of their eggs.[33] And the phenomenon remains a source of fascination to this day. Depending on the species, three sources of heat are used by the megapode family: heat generated by respiring micro-organisms while decomposing organic material; heat associated with volcanism; and solar radiation.

The family Megapodiidae consists of 22 species in seven genera[33] and includes the brush-turkeys *Alectura* and *Aepypodius*, and malleefowl *Leipoa* of Australia; the *Talegalla* of New Guinea; the Maleo *Macrocephalon* of Sulawesi; the Moluccan Megapode *Eulipoa*; and the widely distributed megapodes *Megapodius*.

In Lombok, Wallace was fascinated with the mounds of the Orange-footed Megapode *Megapodius reinwardt* he saw here and there in dense thickets. These, he said, were a great puzzle to strangers: who could possibly have heaped together cart-loads of rubbish in such out-of-the-way places? Such visitors dismissed as 'the wildest romance' the preposterous notion that the mounds were built by birds.

The Orange-footed Megapode, the most widespread of the megapodes, has no time for romance. Strictly monogamous, Mr and Mrs average megapode spend most of their working week constructing or maintaining the biggest mounds of any of the Megapodiidae. The construction takes several weeks and

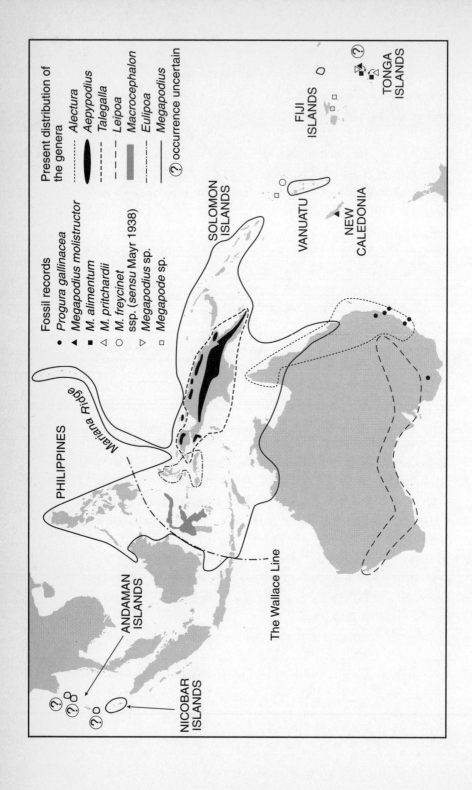

involves the relocation of several tonnes of material. Small hillocks 12 metres across and five metres high are not uncommon. Microbial respiration begins quickly and the mound soon becomes quite hot—enough for cooking eggs, let alone incubation—before stabilising at a level suitable for incubation, at around 32 to 35 degrees Celsius. The 'thermometer-bird' tests the mound by inserting its head into the core.

Maintenance of the mound involves the regular addition and mixing of fresh, damp material. These mounds can be in use for more than 40 years and last for more than more than 1500 years.[33] The pair share mound-building duties. They may even share the mound with other pairs. Being territorial, however, they avoid contact with these pairs and resolve any differences by arguing vociferously in loud, raucous, wailing cries, like neighbours disputing territory over a fence.

Not all megapodes share the Orange-footed Megapode's workaholic trait. The Philippine Megapode *Megapodius cumingii*, for instance, which has managed to disperse to small islands off north-western Borneo, are known to lay their eggs in decomposing garbage. In Brisbane, Queensland, Australian Brush-turkeys *Alectura lathami* commonly take over compost heaps of lawn clippings in suburban gardens.[33]

But perhaps the cleverest use of heat is that of megapodes using geothermal heat—the heat from volcanos—to incubate

---

Map 2. Megapode distribution

An essentially Australian family, megapodes have dispersed into New Guinea and its surrounding islands, through eastern Indonesia to the Philippines. At first appearance, their western dispersal seems to halt at the Wallace Line. A closer look, however, reveals that megapodes have successfully dispersed to offshore islands of Borneo and as far west as the Nicobar Islands, all places which have few carnivores. A noisy, ground-living bird, which attracts attention to itself, megapode distribution seems to be more affected by the presence of carnivores than any biogeographic barrier. (After Jones et al., 1995.)

their eggs. In Sulawesi, the Maleo *Macrocephalon maleo* uses the warmth of volcanically heated streams. Near hot springs or streams they excavate pits in which they lay their eggs. The size and width of the pits are adjusted depending on the distance to the hot water source. Due to the localised nature of these geothermal sites—often not much larger than one or two hectares—each nesting ground attracts many birds. The most extensive megapode nesting site discovered so far is on New Britain, north of Papua New Guinea. On this island, the Melanesian Megapode *Megapodius eremita* lays at only four geothermal sites. The most extensive of these sites attracts an extraordinary 53 000 birds during the June to September laying period.[33]

The strategy of using geothermal heat sources is thought to have evolved after early mound-building megapodes colonised new habitats and discovered the heat literally 'on tap'. From there the megapodes took the step to using solar-heated sand. The Sulawesian Maleo harnesses both the heat of volcanos and solar-powered sand and has done away completely with mound-building. In northern Sulawesi, Wallace built a little hut on a beach where the Maleos were nesting in loose, hot, black, volcanically derived sand. He observed pairs of birds promenading onto the beach, and commented that 'the glossy black and rosy white of the plumage, the helmeted head and elevated tail, like that of the common fowl, give a striking character, which their stately and somewhat sedate walk renders still more remarkable'. Just above the high water mark, he observed the birds scratch a hole about one metre deep in which the female deposited an egg, covered it with about 30 centimetres of sand and walked calmly back into the forest with her mate.

After about 13 days the birds would return to the beach to lay another egg and so on, until several eggs were laid. Wallace shot and dissected many birds. In each of the females he examined, he discovered a large egg and eight or nine pea-sized ones.

Wallace observed that the eggs were 'so large as entirely to fill up the abdominal cavity and with difficulty pass the walls of the pelvis'. With typical humour he also discovered that the large egg completely filled an English tea cup! In truth, the Maleo lays huge eggs, about 230 grams, representing up to 22 per cent of the female body weight.[33] They lay an estimated 8 to 12 eggs per season, which amounts to about 120 to 180 per cent of the female body weight. The largest number of eggs recorded for a megapode was 56 eggs during a single breeding season,[33] more than three times the weight of the female.

The eggshells of Maleo are white or cream and covered with a pinkish-brown powder which give them the brick-red colouration that Wallace noticed. The effort of laying an egg nearly a quarter of its body weight takes its toll on the Maleo and the birds need breaks of about two weeks between laying eggs. The result is a long nesting season, with chicks hatching independently of their siblings. Not surprisingly, the female, when not laying and building, spends most of her time eating, foraging for insects and fruit on the forest floor, the male solicitously offering the occasional food item. In general the long breeding season is limited by rainfall—a soggy mound or pit is not the ideal state for an incubator.

All megapode eggs, as well as being heavy, have a high proportion of yolk, from 48 to 69 per cent (compared with other birds which have a yolk content of 14 to 35 per cent). The amount of time the eggs are incubated varies from 44 to 99 days.[33] Interestingly, the yolk content and length of incubation compares with reptiles. Sea turtles, for instance, have a similar incubation time. In both reptiles and megapodes, the large amount of high-energy yolk and long incubation period results in highly developed hatchlings. And again, like most reptiles and unlike all other birds, young megapodes are completely independent of their parents from the time of emergence.

Wallace, in Sulawesi, recalled an anecdote from a Mr

Duivenboden of Ternate who assured him that little megapodes could fly the very day they were hatched. Mr Duivenboden had first-hand experience, having taken some eggs aboard his schooner at Ternate. They had hatched overnight and in the morning he had hatchlings flying frenetically in a flurry of feathers this way and that across his cabin.

Actually, the task of digging through sometimes tonnes of matted debris or sand is quite an awesome one for tiny megapode chicks. Buried up to a metre below the surface, the chick lies on its back and works in brief bouts of intense activity, scraping away at the material above it. This is followed by long periods of panting recovery. Sometimes the process takes days. Despite this—and as Wallace himself observed—the chicks, once emerged, run for their lives to the safety of the forest. Virtually nothing is known of how megapode chicks survive. Unlike any other bird, they hatch independently of each other, with no parental care, protection, brooding or social environment.

In Maluku, the Moluccas, Wallace discovered an entirely new species of megapode. Named *Megapodius wallacei* by taxonomists of the time, today this Moluccan Megapode is considered to be so distinctive that a new genus has been erected for it and it is now known as *Eulipoa wallacei*. Wallace himself described it as quite distinctive, 'being richly banded with reddish-brown on the back and wings and it differs from other species in its habits'. Like the Maleo, the Moluccan Megapode comes down to the hot, white sands of the beaches of Maluku to deposit its eggs deep in the sand, after which it cleverly disguises its own tracks by making a confusion of other tracks in the vicinity.

An essentially Australian family, megapodes have dispersed into New Guinea and its surrounding islands, through eastern Indonesia to the Philippines. The genus *Megapodius* is found as far west as the Nicobar Islands in the Bay of Bengal and as far east as Tonga in the midst of the Pacific. *Megapodius*, in particular, has a remarkable ability to disperse. Indeed, the Orange-footed Megapode is currently colonising extremely remote islands in the middle of the Banda Sea. The Micronesian Megapode *Megapodius laperouse* is also colonising islands over distances of up to 100 kilometres from another land mass in the Mariana Islands.[33]

Curiously, no megapodes have been found in Bali, Java, Sumatra and the greater part of Borneo, or, indeed, on any of mainland South-East Asia. How can we explain this distribution? The reason appears to be rooted in the fact that megapodes evolved on a continent that was devoid of large numbers of ground-living mammalian carnivores. Where else but carnivore-scarce Gondwana could a bird spend day after day noisily scratching leaves and soil from the forest floor, occasionally diving bodily into a mound to check its temperature and losing all sight of any predators in the immediate surroundings? The fact that the Nicobar Islands have never had any land connection to mainland South-East Asia and hence do not have carnivores explains the survival of the Nicobar Megapode *Megapodius nicobariensis*. The megapodes that have dispersed to northern Borneo are also only found on carnivore-free small islands. In Sulawesi, the endemic Maleo and the Philippine Megapode *Megapodius cumingii* nest in pre-warmed burrows and hence spend less time at these sites than mound-builders. This and the fact that carnivore numbers are limited in Sulawesi explain why megapodes are able to survive here.

Wallace chided natural history writers of his day who firmly held that the habits and instincts of animals were fixed—in this instance, that all birds must exhibit parental affection. Wallace believed that the 'careful consideration of the structure of a species and of the peculiar physical and organic conditions by which it is surrounded, or has been surrounded in past ages, will often, as in this case [of the megapodes], throw much light on the origins of its habits and instincts'.

However, neither physical nor organic conditions could explain to Wallace the absence of prominent Asian birds on Lombok, and the presence of raucous Australian ones. In the midst of the Orient he was unprepared for echoes of Australia. This is because the fauna of Australia differs from the fauna of Asia more than any of the four ancient quarters of the world differ from each other. Australia naturally has no monkeys, no cats or tigers, wolves, bears or hyenas; no deer or antelopes, sheep or oxen; no elephant, horse, squirrel or rabbit. None, in short, of those familiar types of mammals which are met with in every other part of the world. Instead of these, it has marsupials—kangaroos, possums and wombats—and the duck-billed platypus and echidnas. In birds it is almost as peculiar. It has no woodpeckers and no pheasants, familiars which exist in every other part of the world; but instead it has the mound-making brush-turkeys, the honeyeaters, the cockatoos and the brush-tongued lories, which are found nowhere else upon the globe.

In the early part of the nineteenth century, the reality that different places had different floras and faunas—that South America had different animals to those of Africa, for instance—was considered amazing. In Wallace's time, naturalists were only just becoming aware of the immense diversity of floras and faunas that were being discovered all around the world. Charles Lyell, arguably the most influential scientist of the day, referred to zoological and botanical regions as one of the most interesting and remarkable facts established by the advance of modern science. The fact that different regions of the globe were 'peopled' by different organisms was a phenomenon 'extraordinary and unexpected'.[10] It was only in 1858 that naturalists realised that the world could be divided into six general faunal regions: Palaearctic, Nearctic, Neotropical, Ethiopian, Oriental and Australasian.[56] These are the divisions we still use today.

Lyell had emphasised the need for impassable barriers, such as wide oceans and high mountain ranges, to separate the major provinces—for example, the Himalayas which separate the Palaearctic and the Oriental—not a channel 25 kilometres wide. From an evolutionary point of view, a naturalist would not expect an archipelago of physically similar islands to be populated with strikingly different animals on its western and eastern ends, let alone across a narrow channel. But this was the case between Bali and Lombok.

Between Sulawesi and Borneo, Wallace found the differences to be even more striking. In Borneo, across the Makassar Strait, the forests abounded with monkeys of many kinds, wild-cats, civets, otters and squirrels. In Sulawesi, he found few indeed of these, but instead plenty of prehensile-tailed cuscuses; in Borneo, Oriental birds like woodpeckers, barbets, trogons, fruit-thrushes and leaf-thrushes; in Sulawesi, honeyeaters and parrots.

Sulawesi had even more surprises up its spider-shaped sleeves. Its central position in the Indonesian Archipelago, connected by small satellite islets and coral reefs to the Philippines in the north, Kalimantan to the west, the Moluccas to the east and the Lesser Sundas to the south, led him to expect the richness and variety of the whole archipelago to be found there. Instead the facts turned out to be just the reverse. Sulawesi 'was at once the poorest in the number of its species and the most isolated in the character of its productions of all the great islands of Indonesia'. With a land area nearly double that of Java, it had but half Java's number of mammals and land birds. Its 'productions', hemmed in on every side by islands, were either distantly related or were totally unrelated to those islands.

When he recovered from the shock, Wallace realised that Sulawesi's wonderfully rich, peculiar forms were a marvel. The only way he could explain the anomalies and eccentricities was to suggest an origin in remote antiquity, providing isolation and time enough for evolution to have formed the unique wildlife of, arguably, 'the most singular island in the world'.

In January 1858, the year the joint theory of evolution was presented to the Linnean Society of London, a perplexed Wallace wrote a letter to his friend Henry Walter Bates, in London, describing the existence of a near magical faunal boundary. 'In the Archipelago', he wrote, 'there are two distinct faunas rigidly circumscribed, which differ as much as those of South America and Africa and more than those of Europe and North America. Yet there is nothing on the map or on the face of the islands to mark their limits. The boundary line often passes between islands closer than others in the same group.'[43]

In 1863, in London, Wallace read a paper to the Royal Geographical Society, on the geography of the Malay Archipelago. He drew a red line on the map passing down the Makassar Strait. To the west he wrote 'Indo–Malayan region' and to the east he wrote 'Australo–Malayan region'. This became the Wallace Line.[31]

A mysterious line, only 25 kilometres wide, that separates marsupials from tigers and honeyeaters and cockatoos from barbets and trogons, could not fail to appeal to the imagination of the layman. The concept was enthusiastically adopted by early zoogeographers. Some exaggerated it to the point of absurdity, going so far as to suggest that in a single step one went from the present to the Mesozoic of 100 million years ago.

Exaggerated claims that the Line represented a sharp boundary between two different faunal universes resulted, inevitably, in an equally vigorous reaction. Some biogeographers asserted that such a sharp boundary did not exist. Not only was there none where Wallace drew it, but none existed anywhere in the archipelago. Other biogeographers agreed that the Wallace Line was entirely imaginary but, based on new distributional information of anything from earwigs to elephants, drew other lines of their own. Still others, in turn, vigorously defended the Line. The Wallace Line was (and still is) a conceptual trampoline,

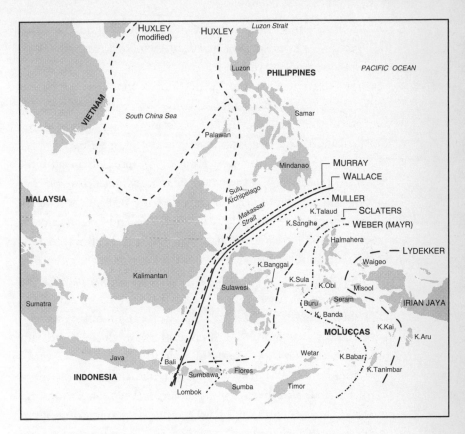

Map 3. Biogeographic lines of Wallacia

The Wallace Line is not the only biogeographic line to be drawn in Indonesia. Salomon Muller drew his line in 1846; Wallace originally defined his line in 1859; Murray's Line was defined in 1866; Thomas Henry Huxley modified Wallace's Line in 1868; Lydekker's Line was drawn in 1896; Sclater's was drawn in 1899; and finally there was Weber's Line of 1902. 'Too many lines' said biogeographer George Gaylord Simpson in apparent exasperation. He likened the whole exercise to an arbitrary game in which boundaries between one region and another must be definite, in order that biogeographers could replace one colour with another on a map.

bouncing biogeographic claim and counter-claim.

Interestingly, the Wallace Line was not the first biogeographical line to be drawn in Indonesia. In 1846, Salomon Muller defined a line based essentially on ecology, pointing out, for example, that arid conditions run eastwards from Java to Tanimbar and affect the vegetation. Muller's Line was similar to the Wallace Line, passing just to the east of it, between Sumbawa and Flores in the Lesser Sundas and then essentially following the same route north through the Makassar Strait. In fact, six distinct lines have been proposed, in addition to the Wallace Line, as regional boundaries in this area.

The Wallace Line, the most famous and most discussed biogeographic boundary in the world, was based essentially on the distribution of birds, especially parrots. Wallace originally intended that the Line run between Lombok and Bali and north through the Makassar Strait between Kalimantan (Borneo) and Sulawesi, turning north-east to the Pacific Ocean between the Philippines and the small Indonesian island groups of Sangihe and Talaud.

Wallace, in fact, vacillated about which region Sulawesi truly belonged to, not committing himself in 1860, although he had done so in 1859 in a letter and did again in 1863, only to reverse his opinion in 1910. Finally he said it would ever remain a matter of opinion whether it be considered Asian or Australian, remarking that 'there is no other example on the globe of an island so closely surrounded by other islands on every side, yet preserving such a marked individuality in its forms of life; while, as regards the special features which characterize its insects, it is, so far as yet known, absolutely unique'.[74]

A second line, Murray's Line of 1866, was based on the distribution of mammals and had a route similar to the Wallace Line. The only difference was that Murray's Line passed between Bali and Java, rather than Bali and Lombok, before continuing north through the Makassar Strait.

It was actually in 1868, 10 years after Wallace wrote his letter to Bates and five years after he presented his paper to the Royal Geographical Society, that the term 'Wallace's Line' was coined. In that year, Thomas Henry Huxley (1825–1895) published a paper based on the distribution of birds. Huxley drew a line passing between Bali and Lombok and then northward through the Makassar Strait. He then continued it further northward between Kalimantan and the Sulu islands and then between Palawan and the Philippines. He called this Wallace's Line. Wallace, however, never accepted this northern extension of his Line and thus it is properly called Huxley's Line.

Lydekker's Line of 1896 corresponds with the edge of the Australian continental shelf. In a sweeping curve, it enfolds the islands of Waigeo, Batanta and Misool, off the Vogelkop (bird's head) Peninsula of Irian Jaya and the Aru islands to the south of Irian Jaya.

In 1899, father and son naturalists Philip Lutley and William Lutley Sclater drew yet another line based on mammalian distributions. It passed between Bali and Lombok and then swung round the east of Sulawesi, through the strait between the Banggai and Sula islands and finally around the extreme north-eastern tip of Halmahera, out into the Pacific.

Finally, there is Weber's Line of 1902. Originally it was based on the distribution of freshwater fish. German biologist Ernst Mayr however, redefined it in 1944 as the line of 'faunal balance': west of the line, the fauna is considered more than 50 per cent of Oriental origin and east of the line more than 50 per cent Australian origin. Ironically this line separated areas which were most similar. It was the antithesis of a faunal break and for this reason was largely disregarded. Essentially the line traces out the boundary of the North Moluccas, running first between Timor and Australia and then northward between the islands of the Babar and Tanimbar groups, west around Buru and Halmahera and finally into the Pacific.

'Too Many Lines', said biogeographer George Gaylord Simpson in 1977, in a paper by that title. He likened the whole exercise to an arbitrary game in which boundaries between one region and another must be definite, in order that biogeographers could replace one colour with another on a map. In the end Simpson refused to play. And of all the faunal lines mentioned, only the Wallace Line has persisted as an important biogeographical concept.

Not that the arguments of biogeographers mattered. The reality was that *something* had happened in the vicinity of the Wallace Line that could not be denied. German zoologist Bernhard Rensch frequently travelled across the strait between Lombok and Bali in the 1920s. On first visiting Bali from Lombok he said, 'What about the animal life? Is it really different from that of Lombok as has been claimed by so many other travellers? A strait which even the smallest bird could cross without difficulties ... And the difference is indeed quite extraordinary! Much more conspicuous than I would have ever imagined. As soon as I entered the woods on a small native trail a whole chorus of strange bird songs greeted me—in fact, among the real songsters there is not a single one with which I was familiar'.[41]

Ornithologist G.A. Lincoln studied birds in Java, Bali, Lombok and Sumbawa, on either side of the Wallace Line, in the 1970s. He found that on Java and Bali the birds were nearly all Oriental and dominated by the Oriental Bulbul. On crossing the Wallace Line from Bali to Lombok the birds were predominately Australian. The honeyeaters, in particular, attracted attention with their noisy calls and squabbling behaviour. Java and Bali had the most similar bird faunas. The least similar adjacent bird faunas were those between Bali and Lombok, astride the Wallace Line.

Zoologist R.C. Raven sailed back and forth across the Makassar Strait from Kalimantan to Sulawesi, from 1912 to 1923, collecting natural history material for American museums. Raven found that while a number of mammalian groups had indeed succeeded in passing the Wallace Line from the west (he was not concerned with movement from the east), a much greater number had been stopped by it. While his data, based on 1935 systematics, is now somewhat out-of-date, the general trend that he observed remains true. For instance, he found that of 59 species of shrew, only 13 crossed. Only two of 21 species of deer in the region had crossed the Wallace Line. Of 56 species of civet cats he identified, only seven species were found east of the Wallace Line. Of 196 species of squirrel, Raven said only 14 were found in Sulawesi, across the Wallace Line.

The specialised Pangolin, which ranges from India and China to Java and northward through Borneo to Palawan and the Philippines, does not naturally cross the Wallace Line. Tapirs, rhinos, flying lemurs and hedgehogs extend only as far east as Java and Borneo.

None of the 16 species of the insectivorous tree shrew of the family Tupaiidae, which extend right to the Wallace Line, had crossed it. The three species of the Asiatic mouse deer, or chevrotain, with its headquarters in the Malayan region, including the Philippines, do not cross the Wallace Line. The Muntjacs, or Barking Deer, also, are limited to the Malayan region. They, too, do not extend east of Borneo.

Only one canid, the Dingo, crosses the Wallace Line. The once widespread Wild Dog of Asia *Cuon alpinus* extends east to Sumatra and Java but no further. The Malayan Sun Bear *Helarctos malayanus* is found in Malaysia, Sumatra and Borneo but not east of the Wallace Line. Of 27 species of Mustelids— weasels, badgers, skunks and otters—Raven identified, none had transgressed the Wallace Line.

Hippopotami and the Giraffe, extinct today in Indonesia, but known from the Pleistocene in Java, did not cross the Wallace Line.

Of the Primates, *Tarsius*, macaques and Man occur to the east of the Wallace Line. The lemuroid Slow Loris extends east as far as Java, Borneo and the Philippines. The leaf-eating monkeys, which range over the Malay region including Borneo, do not transgress the Wallace Line. The highly specialised Proboscis Monkey *Nasalis larvatus* is characteristic of Borneo. Of the anthropoids, the gibbons have nine species in the Malay Peninsula, Java and Sumatra, Borneo and Palawan. The Orangutan is found in Borneo and Sumatra. These anthropoid Primates are of east Asiatic derivation and none have transgressed the Wallace Line.

In Wallace's 1858 letter to Bates, written on thin paper in a rickety rattan hut, he described the existence of a near magical faunal boundary. He had an explanation to account for this extraordinary juxtaposition of two fauna which were poles apart. 'I believe the western part to be a separated portion of continental Asia, the eastern the fragmentary prolongation of a former Pacific continent.' This sentence marks the birth of biogeography. He had tried to explain where species come from, and why they occur where they do. To Wallace, questions of species' origin and species' distribution were inextricably linked. He was the first to realise that to really understand the distribution of species over the face of the earth (and in particular in the Malay Archipelago), one had to appreciate, not only the species' evolutionary history, but also the geological history of the region where they occurred. His Line merely underscored this important principle.

CHAPTER

WHERE WORLDS COLLIDE

> I think geologists are more converted [to the concept of evolution] than simple naturalists because they are more accustomed to reasoning.
>
> Charles Darwin to Alfred Russel Wallace, May 1860.

ALFRED RUSSEL WALLACE had a vivid imagination. So it is no wonder that Lewis Carroll's fantastically imaginative *Alice in Wonderland*, first published in 1896, would became a favourite book in Wallace's old age. Filled with riddles and ridiculous situations, where animals ran helter-skelter in caucus-races which no-one won, pompous blue caterpillars smoked pipes atop magic mushrooms and mock turtles sang of soup, where a smiling Cheshire cat vanished in mid-air leaving only its grin to sail among the trees and where seas were formed of tears, it was a book to delight Wallace's playful and rich imagination. In fact, he imagined the natural world to be no less fabulous.

For instance, to account for the formation of the Malay Archipelago, Wallace imagined the skyward heaving of Atlantic-deep ocean beds. Accompanied by a crescendo of earthquake shocks, volcanic action sent rivers of molten rock and sediment pouring down into the sea from the lands on either side, to build up the continents so that they spread toward each other. Islands rose, Fantasia-like, in the intervening ocean channel, as continents strained toward each other, now connected, now disconnected. In this way, all the islands east of Java, Wallace ima-gined, once formed a part of a former Australian or Pacific continent, although some of them may never have been actually joined to it. This great continent had been shattered into bits over time, Sulawesi possibly marking its furthest western extension.

Wallace reasoned—incorrectly as it happens—that it would be impossible to tell the origins of the various fragments of land in this archipelagic jigsaw puzzle. The animals and plants, however, would reveal clues of their former history. 'Some portion of the upraised land might at different times have had a temporary connection with both continents and would then contain a certain amount of mixture in its living inhabitants [as with Sulawesi] ... Other islands, again, even though in such close proximity as Bali and Lombok, might each exhibit an almost unmixed sample of the productions of the continents of which they had directly or indirectly once formed a part.'

Imaginative as this vision was, it was, nevertheless, conjured within the framework of contemporary geologic theories. These theories, essentially those of Charles Lyell, held that continents were ancient and permanent. For sure, geological dislocations and reconnections, acting slowly enough for living beings to disperse or retreat at an entirely natural pace, were allowed for. But these structural changes acted in only one direction.

According to Lyell and everyone before him, the continents moved—but only up and down. Using this world view to explain the many faunal similarities between distant lands, whole continents had to rise from the ocean depths, forming faunal highways. In this way the romantically named 'Lemuria' rose, streaming from deep below the Indian Ocean, like some ancient Atlantis, to levitate lemurs from Madagascar to India and hence to the Malay Archipelago.

If—as Darwin quipped to Wallace—geologists are more accustomed to reasoning, then surely reason treads where imagination fears to go. Astronomer and geologist Alfred Wegener had no such fear. He was one of the great Arctic explorers, trudging dangerous frozen wastes. He also trod treacherous conceptual paths; of course continents moved vertically, said Wegener, but, he reasoned, they also moved laterally. In fact, they rafted across the face of the globe like gargantuan armadas, sometimes moving closer together, sometimes drifting apart.

Wegener called his theory Continental Drift, announcing it in 1912, a year before Wallace's death. As outrageous as this idea seemed, Wegener had nevertheless compiled an impressive list of evidence supporting continental drift: reconstructions of ancient climates; the evidence of ancient life; the geometrical fit of continental margins on opposite sides of oceanic basins; and the matching of rock successions and truncated geological structures across oceans. Wegener even proposed the name 'Pangaea' for a single supercontinent presumed to exist 180 million years ago. For his troubles, he was virtually shouted off the stage and his work was considered 'utter damned rot'![49] Undeterred, Wegener elucidated his revolutionary theory—the first ever attempt to explain the evolution of the continents—in a book, *The Origin of Continents and Oceans* (in his native German), which was first published in 1915. He revised the fourth edition in 1929, the year before his death.

The debate for and against continental drift was argued with deadly sarcasm—a little like the medieval philosophical controversy regarding Noah's Ark. It was true that there was no plausible mechanism for continental drift at the time and geophysicists trenchantly argued that there was no known physical property of the Earth that would possibly permit such movement.

Proponents of the theory, on the other hand, argued that it was not right to ignore the glaringly obvious supporting geological facts, simply because geophysicists were not yet smart enough to conceive a physical explanation. Other experts, again, disputed the so-called geological facts, calling them instead inferences based on inadequate or inconclusive data. And so it went on.

Wegener didn't let the vicious attacks destroy him. That was left to the Arctic ice. In 1930, he organised an international scientific team to investigate weather conditions on the Greenland icesheet. This was partly in anticipation of future commercial flights between America and Europe, Greenland being on the shortest route. Three research parties were established on Greenland—one in the middle of the icesheet. In 1930, it took three expeditions, each with three dog sleds, to equip the central station, 400 kilometres from both the east and west coasts. Wegener arrived in October with the final supplies for the wintering party.

On 1 November, Wegener celebrated his fiftieth birthday in the middle of the Greenland icesheets. Then, with temperatures dropping to minus 50 degrees Celsius, he and one companion departed for the coast. It was a grim race through the darkening winter and they were lashed with icy wind and drifting snow. They did not make it. Wegener's body was found buried in the snow. His companion and the sled were never found.

WHERE WORLDS COLLIDE

A major objective of Wegener had been to test a new technique known as echo-sounding, which would determine the thickness of the icesheet over Greenland. The idea was that a surface explosion would send waves down through the ice, which would be reflected from the rock below the icesheet and recorded at the surface. From the known speed of travel of the waves and the measured time of travel, the thickness could be calculated. He was never to realise the success of this technique, which, poignantly, would be used not many years later to investigate ocean floors and be instrumental in the revival of interest in continental drift and plate tectonics.

In 1949, conventional ideas surrounding the formation of the Indonesian Archipelago invoked the mushrooming of gargantuan mountain roots beneath the islands. Their billowing growth would eventually weld together the Asian and Australian mainlands. These mountain roots, like colossal, underground granitic fungi, were thought to be the convulsive products of systems of mountain-sized energy waves moving through the Earth's crust, which were in turn driven by the energy generated by physico-chemical processes within the rock. Dutch geologist Reinout W. van Bemmelen, reflecting the ideas of his time, said that 'the majority of the geologists, familiar with the geology of this part of the Earth's surface, reject the hypothesis of a drifting Australian continent'.[66]

Wegener, as it turns out, had the last laugh. Today we know that the continents have drifted across the surface of the globe through much of Earth history. Supercontinents have rifted apart and oceans have opened, expanded and disappeared again as continents collided.

According to plate tectonic theory, the surface of the Earth is covered by a series of relatively thin but rigid, shell-like plates (see Map 4, page 46). These plates of rock slide over the Earth's surface, driven by gigantic convection cells of hot, solid rock material within the Earth (a bit like the oft-used analogy of scum

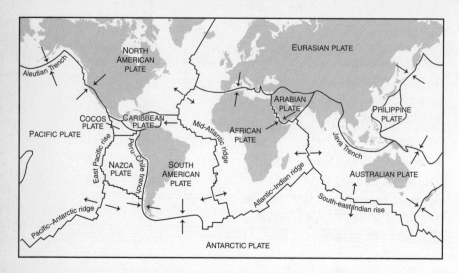

Map 4. The Earth's tectonic plates

The surface of the Earth is covered by a series of relatively thin but rigid, shell-like plates. There are 12 major plates and numerous minor ones.

These plates of rock slide over the Earth's surface driven by gigantic convection cells of hot semi-solid rock material.

on a slowly simmering pot of thick pea soup), grinding against each other, breaking apart and colliding together, apparently at random.

Like children with their heads over the pot, geologists who study plate tectonics are mesmerised by the movement of the scum. Predicting collisions between bits and pieces is difficult enough. But imagine first looking at a simmering pot of soup after an hour and then predicting what the original configuration of scum was! It is possible to do it, but geologists have had to throw in great lumps of imagination, with that reasoning generously attributed to them by Darwin. For Indonesia, the outcome of putting the pieces together has been extraordinary, to say the least.

Until recently, most continental reconstructions have shown

the supercontinent of Pangaea as a huge horseshoe-shaped clustering of continents, with the Tethys 'Sea' forming a vast, empty ocean twice the size of the Pacific, inside the horseshoe. South-East Asia was always considered an integral part of Eurasia at the top of the horseshoe and Australia with India and Africa in the southern hemisphere.

Recent studies however, have stirred the pot (so to speak) so that now it is clear that large parts, if not all, of South-East Asia had their origins in Gondwana—off the northern and north-western margin of Australia (see Map 5 below).[45, 46, 47, 48] Australia, at the time, was the easterly tip of a continent-sized peninsula of Gondwana. Over the past 500 million years, not only pieces of Indonesia close to Australia, but also parts of

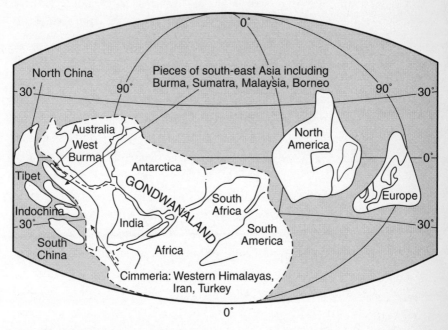

Map 5. Ordovician Around 500 m.y.a.

The latest geological evidence suggests that large parts, if not all, of South-East Asia had their origins in Gondwana, lying off the easterly tip of a continent-sized peninsula which contained Australia. (After Metcalfe, 1993.)

Borneo and Sumatra, all of Malaysia and Indochina—even South China, North China and Tibet—have rifted from the northern and north-western margins of Australia and drifted across the ancient Tethys ocean, to dock against pieces of today's Siberia and Mongolia. Wallace would have been agog!

The evidence for this literally earth-shattering revelation comes from a number of fascinating sources including stratigraphic, palaeontological and palaeomagnetic data.

Rocks provide a window into ancient environments. For instance, sandstones often indicate sandy marine environments and conglomerates indicate a turbulent environment—perhaps a fast-running river dumping rounded, even-sized pebbles at the base of a mountain range. Rock layers containing angular pieces of vastly different size, such as house-sized boulders among small stones and rocks, indicate the action of a glacier plucking up rocks at random and dumping them where it melts.

The different layers in rock faces are, in many respects, fossil environments. The layers of rocks are stratified, meaning that the oldest rocks will be buried by the youngest and so on. But when rocks are violently forced against each other, such as at the front-line of a continental collision, they act like geological toothpaste. Here layers are crumpled, twisted and folded, sometimes end-over-end. Using this sort of stratigraphic evidence it is possible to match rocks which have had the same history even if they are on opposite sides of the planet.

Continental break-up, in particular, follows a clear sequence of events, which itself can be trapped as layers of rock. Pre-breakup, break-up and post-breakup of a rift–drift sequence, for instance, have unique characteristics. The pre-breakup stage is characterised by a long period, maybe 50 million years or more, of rift valley tectonics—like the rift valley of East Africa. Rivers and rain heap sediment into the rift valley which, later, form recognisable rock-types or layers in a section of rock. The break-up stage involves major faulting, with huge slabs of rock slipping and sliding against each other causing earthquakes, uplift, major erosion and, consequently, volcanism. The sundering of continents is accompanied by an upwelling of lava—essentially new seafloor—which squeezes between the continents and pushes them apart. The continental margins slump, and shallow sea environments form, followed by open marine environments with low sedimentation rates.

Fossils of plants and animals are a more obvious way of matching now-sundered continents. This is because ancient environments, like today's environments, were regional—for example, the thinly forested Kimberley region of north-western Australia in comparison with the jungles of Borneo—and harboured site-specific suites of species. These can be matched, even if they now lie in rocks on different continents in different hemispheres.

Rocks not only entomb ancient plants and animals but also ancient magnetic fields. It is an extraordinary fact that about every million years the Earth's magnetic field reverses its polarity. For some unknown reason, the magnetic North Pole becomes the magnetic South Pole and the magnetic South Pole becomes the North. (The geographical poles, themselves, do not change, just the direction of the magnetic force lines.) Certain rocks, such as solidifying lava, can take on the magnetic field of the time.

Lava emerging from a volcano solidifies at 900 degrees Celsius, at which point the rock consists of a mesh of small interlocking crystals. Among these minerals are some, such as magnetite, which can be magnetised. At 900 degrees Celsius the atoms composing the magnetite are vibrating too much to be affected by the Earth's magnetic field. As the rock cools, however, the Earth's magnetic field begins to exert itself on the magnetite, and small atomic groups within the magnetite begin to line up parallel to one another in the direction of the Earth's magnetic lines. By the time the temperature has dropped to 450 degrees Celsius, the magnetic field is locked in and each small mineral has become a small magnet, with a polarity parallel to the Earth's current magnetic field.

This record of the magnetic field at the time and place of formation of a rock persists, even if the rock drifts off to another place on the globe where the existing magnetic field differs from the record trapped in the rock. Not only do the tiny minerals align themselves parallel to the Earth's magnetic field, they also dip or tilt towards the nearest Pole. In effect they behave exactly like miniature compass needles.

If this rock can be accurately dated using modern methods of timing radioactive decay, or stratigraphic evidence indicating its position in relation to rocks of known age, then it is possible to date the specific magnetic field of that rock when it was formed. By knowing this information from many rocks of different ages on a continent, by integrating stratigraphic, palaeontological and palaeomagnetic data, it is possible to unravel the wander paths of continents.

Using this information, geologists have recently matched 500-million-year-old rocks from north-western Australia with rocks now in South China, Burma, Thailand, Malaysia and north-western Sumatra. If you could fast-rewind the last 500 million years, you would witness the dismemberment of Asia as huge chunks split off, and meandered, drifted and collided in mid-ocean and rotated across the face of the globe, to glide into dock along the north-western margin of Australia which lay in close proximity to Iran and Himalayan-India (see Map 5, page 47).

Five hundred million years ago (even before formation of the vast supercontinent of Pangaea, in which all the continents on Earth were temporarily joined around 240 million years ago), all these places formed a huge peninsula of a vast supercontinent called Gondwana, which more-or-less straddled the equator. Gondwana was surrounded by the mild waters of the ancient

Tethys ocean—the Pacific, Indian and Atlantic Oceans were not even a glimmer between close-fitting land masses. Even Antarctica was nowhere near the South Pole. Ancient coral reefs and salt deposits, fossilised in rocks of the time, indicate that hot, tropical climates reigned.

Off the coast of this vast, island-studded peninsula, the tropical seas swarmed with animals which were endemic to the region. Certain invertebrates—specific nautiloids, a Gondwanan snail *Peelerophon oehlerti* and Gondwanan trilobites—flourished only in these waters. They lived out their lives in water, and in dying, drifted to the soft sediment of the sea floor. Today, as fossils, these animals are found as far afield as Tibet, Burma, Thailand and Malaysia.

Between 440–395 million years ago, North China, South China, Indochina and eastern Malaysia rifted from Gondwana and began their journey across the Tethys ocean. They travelled at different paces. Slabs of South China, for instance, drifted alone for perhaps 50 million years during the Devonian period, from 395–345 million years ago. This was the period in evolutionary history when fish were all the rage and South China developed a suite of fish and other animals which were highly endemic (see Map 6, page 52).

The collision of South China with Indochina, in the middle of the Tethys, caused extensive deformation of rock along a belt several hundred kilometres long (see Map 7, page 53). This belt, now a northwest–southeast slash in northern Vietnam, is the 300-million-year-old suture between South China and Indochina.

Back on Gondwana, more microcontinental fragments were being shed. At around 260 million years ago, a rather narrow continent, several thousand kilometres long, called 'Cimmeria', was peeling off the north-western coast of Australia, northern India and northern Africa, which were connected to each other as part of the great supercontinent of Pangaea (see Map 8, page

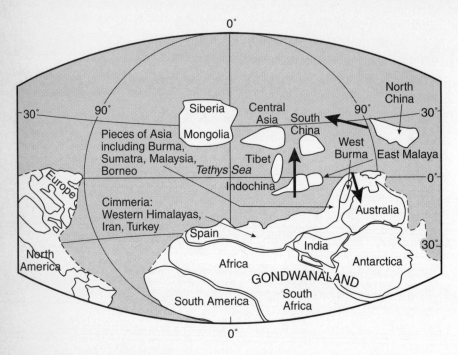

Map 6. Devonian Around 400 m.y.a.

Roughly around 400 million years ago, North China, South China, Indochina and eastern Malaysia rifted from Gondwana and began their northerly journey across the Tethys Sea. (After Metcalfe, 1993.) The heavy arrows indicate continental movement.

54). Cimmeria moved swiftly through the Tethys, rather like a windscreen wiper. As it drifted it began to break-up, the western end eventually forming parts of Iran and Turkey. The eastern end, containing pieces of Burma, Thailand, Malaysia and Sumatra collided, around 220 million years ago, with South China and Indochina, still loitering in the middle of the Tethys (see Map 9, page 55).

This scramble of terranes, situated just north of the equator, finally collided with North China and Eurasia, which were part of the northern supercontinent of Laurasia, about 190 million years ago. The collision was so powerful that it slammed North

WHERE WORLDS COLLIDE 53

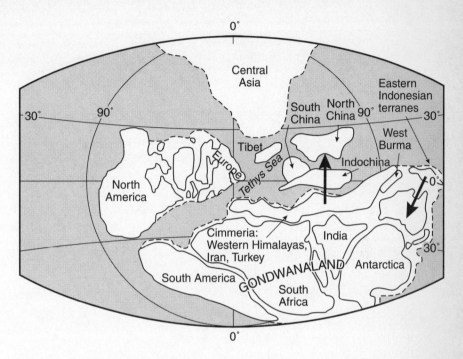

Map 7. Carboniferous Around 300 m.y.a.

Around 300 million years ago, South China collided with Indochina in the middle of the Tethys. A northwest–southeast slash in today's northern Vietnam is evidence of this suture. (After Smith et al., 1981 and Metcalfe, 1993.)

China into a spectacular anticlockwise spin, threw up mountains like heaped scar tissue at the suture line and resulted in the consolidation of the core of South-East Asia.

Meanwhile Australia continued to peel. During the Jurassic, from 190–136 million years ago, a second continental sliver rifted from Australian Gondwana (see Map 10, page 56). This continental sliver drifted north as new sea floor was formed along a rift zone spreading from the Kimberley Region of Western Australia, up into Papua New Guinea and then arching back into eastern Australia. Many of the structural features in the basement of New Guinea were formed during this break-up.

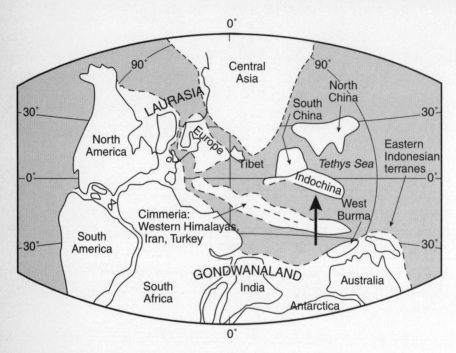

Map 8. Middle to late Permian Around 260 m.y.a.

Gondwana continued to shed microcontinental fragments. At around 260 million years ago, a narrow continent several thousand kilometres long, called 'Cimmeria', peeled off the north-western coast of Australia, northern India and northern Africa, which were connected to each other as part of Pangaea. (After Smith et al., 1981 and Metcalfe, 1993.)

Australia, at the time, was in high latitudes, with New Guinea as far as 40 degrees south.

Along this vast rift valley new sea floor oozed out and began to spread. Sea water poured into the deepening valley, first as a shallow marine embayment and then, crashing in, as open ocean—thus heralding the birth of the proto-Indian Ocean which was linked across the top of Australia with the proto-Pacific. If a sailor could have navigated this open ocean in the late Jurassic, she would have been baffled by a screen of continents and microcontinents.

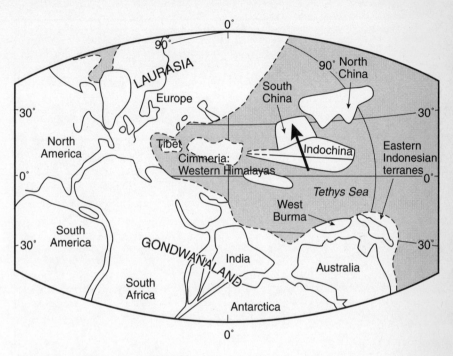

Map 9. Late Triassic Around 220 m.y.a.

As Cimmeria moved through the Tethys, rather like a windscreen wiper, it began to break-up. The western end eventually formed parts of Iran and Turkey, while the eastern end contained pieces of Burma, Thailand, Malaysia and Sumatra. This map shows that the collision between South China and pieces of South-East Asia, with North China, to create the core of South-East Asia, is imminent. (After Smith et al., 1981 and Metcalfe, 1993.)

Closest to Australia was a chunk of a dismembered proto-New Guinea—essentially the southern half without the Vogelkop (bird's head) and its Peninsula. Nearby, to the north, at 30 degrees latitude, was the rest of fragmented New Guinea. Eastern Sulawesi, Obi–Bacan, Banggai–Sula, Buton, Buru–Seram and a chunk of western Irian Jaya would eventually align to form the exotic Spice Islands of eastern Indonesia. Timor and the Tanimbars lay more or less in the same relationship to Australia as they are today (see Map 11, page 57).

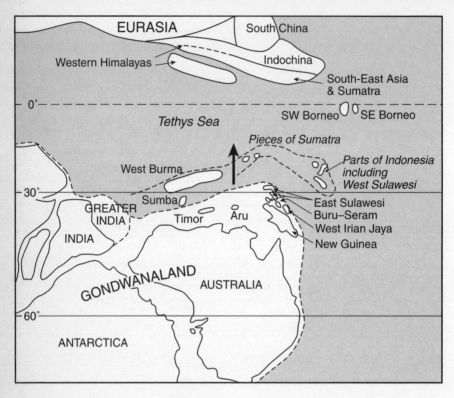

Map 10. Late Jurassic Around 160 m.y.a.

While South-East Asia was consolidating, Australia continued to peel. A second continental sliver rifted from an arc across northern Australia from the Kimberley region, up into Papua New Guinea and arching back into eastern Australia. Many structural features in the basement of New Guinea were formed during this break-up. (After Smith et al., 1981, Audley-Charles et al., 1988 and Metcalfe, 1993.)

In an arc to the north of the Spice Island fragments and New Guinea, lay Sumba, western Burma, western Sulawesi, Nusa Tenggara (the Lesser Sundas) and various other Indonesian terranes which were caught in a tectonic doldrum, just south of the equator.

Further north again, hugging the equator, were two chunks of south-western Borneo, which had probably been carried away

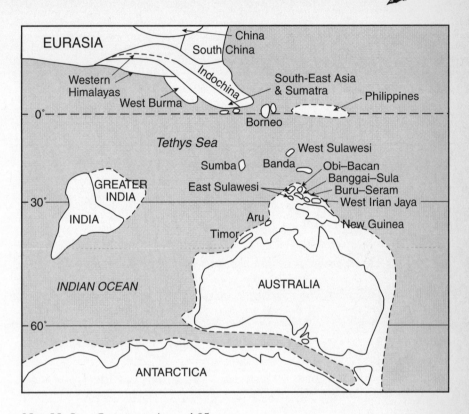

Map 11. Late Cretaceous Around 85 m.y.a.

At this period, a screen of continents and microcontinents separated Australia from South-East Asia. Chunks of New Guinea, eastern Indonesia and bits and pieces of South-East Asia littered the seaways. India, also, broke away from Gondwana, opening up a new ocean, the Indian Ocean. (After Smith et al., 1981, Audley-Charles et al., 1988 and Metcalfe, 1993.)

from Gondwana with Indochina (between 440–395 million years ago), then later split from that embrace to drift back down toward the equator. They collided with each other some time during the Cretaceous (100–65 million years ago) (see Map 11 above).

Rifting was all the go in this period. India broke its long relationship with Gondwana. Rapid sea floor spreading sent it shunting toward the Eurasian continent between 84–55 million years ago, opening the Indian Ocean in its wake and, in front of it, violently shoving the ancient Tethys underneath Eurasia. The Himalayas heaped skyward in reaction. Western Burma and bits of Sumatra got caught up in the Indian slipstream and were cemented into South-East Asia.

Australia, too, rifted from Antarctica and began its northward drift. She caused havoc as she stormed away from Antarctica, snapping the leash of land that held back the bitingly cold circumpolar current which engulfed the lone remnant of Gondwana and left it sheathed in ice.

This continental divorce was complete by 40 million years ago, Australia's crumpling northern apron sweeping before it a collage of crustal fragments. Interactions between the Indo–Australian, Pacific and Eurasian plates were becoming intense. The island of Java was squeezed from the sea floor somewhere near the location of Timor today and began shunting toward South-East Asia.

Bits and pieces of south-eastern China and Indochina were breaking off and when a trench opened up the South China Sea around 50 million years ago, these moved south to collide with south-western Borneo around 25 million years ago. Map 12, on page 59, shows these events.

Stress lines in the form of major transcurrent faults resulted from the westward-moving Pacific plate and the northward-moving Australian plate. These acted like highways, along which Gondwanan-derived blocks of western Sulawesi, Borneo and the rocks which underlie Nusa Tenggara began to slide into their present position around 35 million years ago, followed by eastern Sulawesi, Buton, Banggai–Sula and Obi–Bacan, which reached their current positions in the Maluku province of eastern Indonesia three million years ago.

Other oceanic islands, not caught up in this westward drift, were smeared onto the advancing northern bumper (essentially

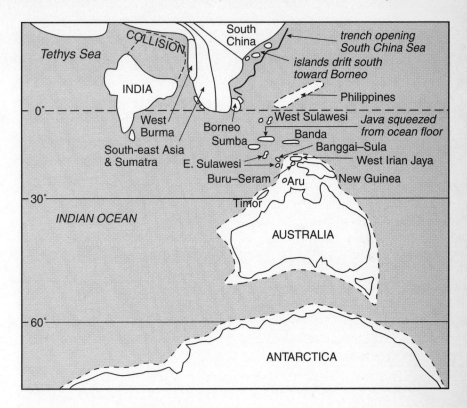

Map 12. Late Eocene Around 50 m.y.a.
Rifting was all the go, Australia finally getting into the act around 50 million years ago. When she stormed away from Antarctica she snapped the leash of land that held back the bitingly cold circumpolar current which engulfed the lone remnant of Gondwana and left it sheathed in ice. (After Smith et al., 1981, Audley-Charles et al., 1988 and Metcalfe, 1993.)

southern New Guinea) of the Australian plate, around 10 million years ago. Continued left-lateral shearing of the Pacific plate across New Guinea caused rapid uplift of the New Guinea Highlands from five million years ago and dragged the Vogelkop, Sulawesi, Seram and the rest of the Banda arc into a megafold in a region of intense biogeographic transfer.[68] Maps 13 to 17 show this complex movement from 20 million years ago, until 3 million years ago.

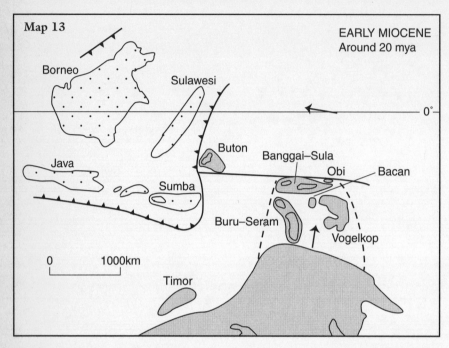

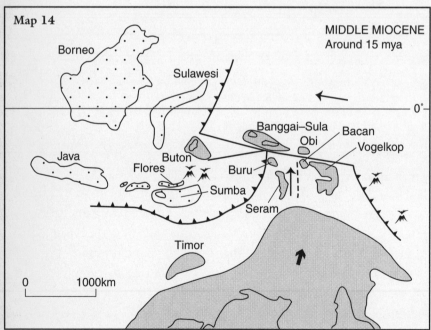

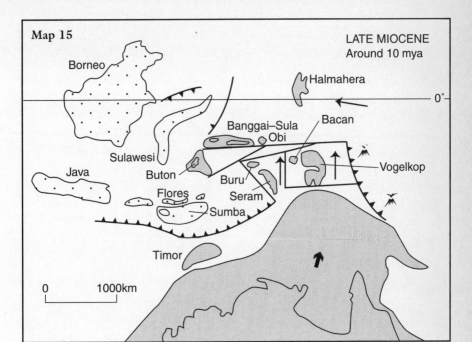

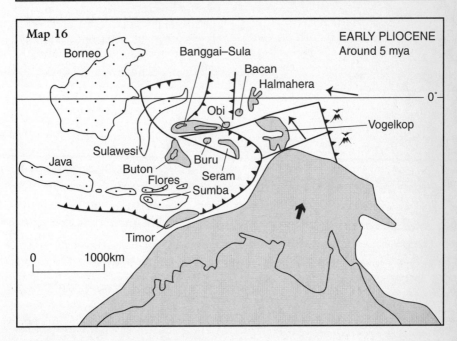

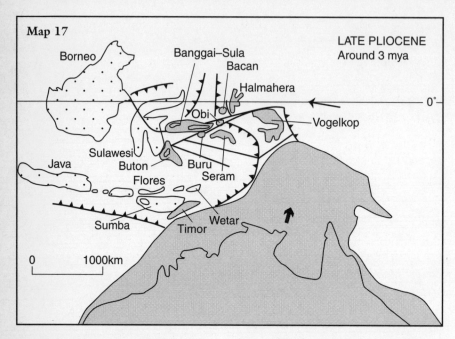

Maps 13–17. This series of maps provide an indication of the complex microcontinental shuffle that resulted from the interaction of the westward-moving Pacific plate and the northward-moving Australian plate. Stress lines from this collision are expressed as faults. These act as highways along which microcontinents move. (After Burrett et al., 1991.)

Does the assembly of this breathtaking global jigsaw puzzle fit in with Wallace's observations? And does it provide any supporting evidence for the Wallace Line?

Incredibly, one of Wallace's greatest mysteries is resolved. Sulawesi and Borneo seem never to have been connected. In fact Sulawesi has ever been an island. Wallace, himself, had suggested the possibility of a permanent separation but, in his day, such a thought was treated with scepticism since all islands were considered to have been connected to a continent at one time or another. Ironically, biological confirmation of the fact comes not from land animals, but from fish.

Freshwater fish of the world are generally divided into two

main categories depending on their ability to survive in salty water. Primary division families are those families which are, and always have been, strictly intolerant of salt water. Secondary division families are those that now live in fresh water but are able to tolerate sea water for a short period. Their evolution is rooted in the sea. The natural occurrence on an island of primary division freshwater fish indicates that some land connection must have existed in the past between the island and the mainland. By definition, the fish could not otherwise have got to the island. The numbers of primary division species on Sumatra, Java and Borneo are 232, 104 and 340 respectively.[39] Sulawesi, strikingly, has no primary division fish, though it does have some of the most remarkable assemblages of fishes in South-East Asia. This magic isle will be the focus of Chapter eight.

Wallace had always maintained that major biogeographic regions of the earth correspond to ancient, long-lasting barriers to the diffusion of species. The coherence of Sundaland—as evidenced by zoologist Raven's broad brush-stroke picture of Asian mammalian distribution across the Makassar Strait—indeed has its roots in the 35-million-year-old union of Sumatra, Borneo and Java. Like a giant moat, the Makassar Strait, the most robust section of the Wallace Line, has always separated these from Sulawesi and islands to the east. The distribution of plants, butterflies and birds also reflect this barrier; the butterflies, birds and plants of Sulawesi, for instance, are more like those of Maluku and Nusa Tenggara than they are of those on nearby Borneo.

Of Maluku, in between Sulawesi and Australia–New Guinea, Wallace observed 'that the Moluccan fauna has been almost entirely derived from that of New Guinea'. Plate tectonics confirm that while the islands of Maluku have a rather chaotic geological history, their complicated relationships are all with New Guinea. So far, so good.

Wallace had said that to appreciate species' evolutionary history, one had to understand the geological history of the region where they occurred. And he was right. Modern-day biogeographers, aware of plate tectonic reconstructions, continue to search for common biogeographic patterns among taxa, to fine-tune predictions of geological and climatic changes. They assume that animal and plant groups that have gone through the same geological and climatic events should largely reflect these events in their distributions.

Groups of organisms that evolved in different periods will, in turn, show different patterns of area relationships. But the real problem for biogeographers is that different groups of organisms, forming an ecosystem of an area, will almost certainly have responded differently to the same sequence of geographic, climatic and ecological changes. The first step in the search for general biogeographic patterns, then, is the search for an appropriate animal or plant group that will unambiguously reveal past geologic and climatic changes.

In a story of drifting armada-like continents and magical faunal boundaries, which group could be more appropriate than 'looking-glass' insects to provide a finely tuned mirror to the scene: they are ancient, have a worldwide distribution, are highly endemic and are poor dispersers. As such, they mirror geological history.

Cicadas are singing insects which grow up to six or seven centimetres long. Beloved of children who, in Australia, have given them fairytale names like the greengrocer, yellow Monday, double drummer and the floury baker, they are part of an old order of animals which evolved around 280 million years ago. At this time they could diversify easily around a world dominated by one supercontinent. Their evolution was fuelled by the emergence of flowering plants at around 130 million years ago and with them they radiated extensively. There are now around 1200 species of cicada worldwide.

Cicadas are a perfect model of the biogeographic principle, championed by Wallace, that the distribution of animals and plants in space echoes their distribution in time. This principle predicts that higher taxa such as classes, orders and families are generally spread across the globe. As a family, cicadas do indeed have this cosmopolitan distribution, reflecting its ancient evolution. Genera, and particularly species, should have more restricted distributions, reflecting geologically more recent evolution in a world of changing continental and climatic configurations. Of the 38 Australian genera of cicadas, for instance, 28 are endemic. At the generic level then, cicadas reveal distribution patterns, restricted to large regions of the planet, such as the Australian region. Generic distributions, however, can still echo old patterns: in Australia at least one of the non-endemic genera is well represented in North America, a distribution which echoes an old, pre-Gondwanan pattern. Evolution to species level is a more recent event still. In Australia this is reflected in the fact that all but four of the 250 species are endemic. These species show restricted geographical distributions within Australia, with closely allied species being found geographically near to each other.

Of course, not all animals and plants behave in such a model fashion for biogeographers. In the first place, many groups of animals and plants had not evolved in time to track wayward continents. Others have not diversified enough—there are not enough of them—to track distribution patterns since their evolution from a single ancestor. For instance it is not possible to determine the wander paths of the continents by looking at distribution patterns of, say, elephants, because the family only evolved about 40 million years ago and there are currently only two extant species; hardly a statistically significant number. (Though elephants can tell us something about more recent biogeographical patterns, as we shall see in Chapter five.) Other taxa, again, have widespread distributions, such as seabirds and some rats (and elephants). These animals just fly, raft, or swim over barriers that would stop any self-respecting cicada.

Cicadas are less flighty than butterflies (another group that has been studied) and many other insects. Being rather chunky they are also less likely to be transported elsewhere by birds or wind. That cicadas are homebodies is also indicated by their long subterranean nymphal life lasting from two to six years, but even 17 years has been reported.[18] Cicadas, therefore, are poor colonists. Those that live in an area are likely to be endemic to that area. Their distribution patterns, are also likely to mirror the complex geological history of the region.

Not surprisingly, Oriental cicadas show a distinct line of demarcation along the Makassar Strait section of Wallace's Line. Here, most Oriental genera reach their eastern limits. As expected, endemism of the cicadas on Sulawesi is very strong, even extending to the generic level, a sure indication of a long period of subaerial (above the sea) isolation. As far as cicada relationships go, however, the geographic entity of Sulawesi includes those islands of the nearby Banggai and Sangihe island groups, to the east and north respectively.

Unexpectedly, however, while Sulawesian cicadas show old (at the generic level) relationships with Maluku and New Guinea, they show no such relationships with Australia.

To understand this we need to look more closely at events around 10 million years ago when Australia swept up an arc of oceanic islands to form New Guinea. These islands became northern New Guinea and their largely Asian-derived biota was spliced onto southern New Guinea with its Australian biota. The collision raised the New Guinean highlands. The Vogelkop Peninsula rotated into its present position later, at around 3 million years ago (see Maps 15 to 17, pages 61–62). The island of New Guinea is therefore a geological composite of recent origin, a message fully reinforced by cicadas.

In fact, cicadas do not yet recognise the island of New

Guinea as a geographical entity. Instead they recognise the islands of Maluku and northern New Guinea as one endemic area. This endemic area also includes the Bismarck Archipelago and Admiralty Islands, Solomon Islands, Vanuatu, even Fiji and Tonga. Across this area, which represents the pre-collisional Melanesian oceanic arc, cicadas are closely related at a high level. Cicadas from southern New Guinea are distinct and unrelated to these Melanesian forms. Indeed, only recently have cicadas dispersed from northern New Guinea to the newly formed central New Guinea and vice versa. As yet, no Australian cicadas have evolved to occupy habitats outside of New Guinea and few Melanesian cicadas have evolved to occupy habitats in Australia.

Embedded within this high ranking Maluku–northern New Guinea area of endemism are lower ranking areas of endemism.[18] These more recent relationships involve species or subspecies which have become endemic to parts of the area. Uncannily, the suspected close proximity of Seram, Buru and Sula three million years ago (see Map 17, page 62) is mirrored in the fact that three of the four southern Maluku endemics are found on these three islands.[18] Also remarkable is the fact that some Maluku cicadas show a relationship with northern New Guinea including the Vogelkop and some show a relationship only with northern New Guinea, apparently jumping over the Vogelkop. Possibly this latter relationship was established before three million years ago when the islands of Maluku and northern New Guinea were adjacent—before the Vogelkop swung into its current place.

The Vogelkop Peninsula itself comprises an endemic area of lower order with species from three widespread genera evolving there. Cicadas of central New Guinea, literally the newest area on the map, are rapidly diversifying, so far mainly at the subspecies level, in the many habitats created by the geologically unstable central mountain ranges.

Plate tectonics, then, go at least part of the way toward

unravelling the complex biogeography of Wallace's Malay Archipelago. But what of the biogeographic barrier between Bali and Lombok, the southern section of the Wallace Line? Plate tectonic reconstructions do not suggest a tectonically derived barrier between Java's offshore island, Bali, and Lombok, the western-most island of Nusa Tenggara (the Lesser Sundas). In fact, Bali probably did not even exist as dry land until volcanic activity heaved it above the sea around three million years ago. Other islands of Nusa Tenggara, similarly, have a history of having been built by volcanic action upon ancient, but largely marine, rocks.

Such volcanism, caused by the collision of the Australian plate with the Eurasian plate, is extremely destructive. Before three million years ago, therefore, it is likely that there has been no easy dispersal route along the chain of islands from Lombok to Timor in either direction. And since then the spread of animals and plants has been hindered by fire and brimstone. While plate tectonic reconstruction itself does not help to explain the elusive nature of the biogeographic barrier between Bali and Lombok, volcanos, the explosive manifestation of plate collision, may be a start.

CHAPTER 4

FIRE-SPITTING MOUNTAINS

> Volcanoes assail the senses. They are beautiful in repose and awesome in eruption; they hiss and roar, they smell of brimstone. Their heat warms, their fires consume; they are the homes of gods and goddesses. Volcanoes are described in words and pictures, but they must be experienced to be known. Their roots reach deep inside the Earth; their products are scattered in the sky.
>
> Robert and Barbara Decker. Volcanoes, *1981. Used with permission of W.H. Freeman & Company.*

THE MALAY ARCHIPELAGO, today's Indonesia, is a nation of volcanos. The great sweep of the Sunda arc, over 3000 kilometres from north-western Sumatra to the Banda Sea in the east, results from the collision of the largely continental Eurasian and largely oceanic Indo–Australian tectonic plates. The Australian plate, in the vicinity of the Sunda arc, is moving north-eastward at around seven to eight centimetres per year.[57] Being a thinner oceanic plate, the

Indo–Australian plate is pushed below the thicker, but less dense, continental crust of the Asian continent and this great zone of collision has produced the earth's most mighty chain of volcanos (see Map 4, page 46).

At the eastern end of this chain of cauldrons, in the Banda Sea region of eastern Indonesia, is a confusing swirl of volcanos. These mark a tectonically chaotic region of converging plate fragments which form multiple collision and subduction zones, each edged with volcanos. Elsewhere on the planet, such collision zones are separated by thousands of kilometres. In eastern Indonesia, however, the Sulawesi–Sangihe volcanic arc and the arc of volcanos on the west of Halmahera, trace two different subduction zones. They are separated by a mere 130 kilometres. The nearby Banda volcano marks yet a different subduction zone—where the Pacific Ocean crust is being subducted westward under the Eurasian plate.

These tiny Banda isles are perhaps the most beautiful of the islands of Maluku. Rising seemingly out of nowhere, from the five-kilometre-deep Banda Sea, the Banda islands cover barely 50 square kilometres of dry land—specks on the world map. Yet in the Middle Ages this tiny island group produced the entire world supply of nutmeg and mace. Worth their weight in gold, these spices fuelled Europe's Age of Discovery and the Bandas, along with Ternate and Tidore off the western coast of Halmahera, became known as the fabled 'Spice Islands'.

In medieval Europe, well-organised and armed Arab traders kept a stranglehold on the spice trade, offering only a trickle for which the Europeans paid whatever was asked. Portugal was the first European country to begin a series of epic voyages in search of the Spice Islands for itself. They sent Vasco da Gama and in 1498 he rounded the Cape of Good Hope and made his landfall in India. The thin edge of the wedge, from here it took the Portuguese only 14 years to find the route to Banda and in

1512 a small flotilla of Portuguese vessels made landfall on the Banda islands. In 1596, Cornelis de Houtman pioneered the Dutch route to the Spice Islands, followed by other well-financed merchants. By 1602, the Vereenigde Oostindische Compagnie or V.O.C. (United East Indies Company), a conglomerate of Dutch merchants, was formed and they held the monopoly on the precious spices for some 200 years.

In December 1857, a Dutch mail-steamer travelling at no more than 10 kilometres per hour chugged from Makassar (near today's Ujung Pandang in southern Sulawesi) to Banda. Alfred Russel Wallace was enjoying the trip immensely. At 6.00 a.m. he would take tea or coffee before a breakfast of tea, eggs and sardines. At 10.00, Madeira, gin and bitters were brought on deck as a whet for the substantial 11 o'clock breakfast (which differed from dinner only in the absence of soup). At 3.00, more tea and coffee. At 5.00, more bitters followed by a good dinner with beer and claret. A final service of tea and coffee at 8.00 p.m. concluded the day. Between whiles there was beer and soda water on demand. Wallace, tongue in cheek, called these, 'little gastronomic excitements to while away the tedium of a sea voyage'.

In due course, the leisurely journey reached the lusciously vegetated Banda islands. The epitomy of paradise, the three main islands of the minuscule Banda group enclose a harbour which, to this day, has water so transparent that living corals and even the minutest objects are plainly seen on the volcanic sand at depths of tens of metres. The Banda islands are, in fact, the aerial tip of a huge volcano. Indeed, the curved edge of the largest island, Banda Besar, traces out the rim of a massive sunken volcanic caldera or crater. Rising out of the sea at the centre of this caldera is the perfect cone of the Banda Gunung Api, or 'Fire Mountain'.

Wallace watched the columns of smoke issuing from the continuously active crevices of Banda Api uniting as they rose to form a cloud above the volcano. The impressive awfulness and grandeur of the perfectly conical volcano did not fail to impress him. A volcano was, after all, a fact opposed to his mass of experience in a comparatively benign English countryside. In 1857, volcanos were considered a rarity among European scientists, with less than 200 so far known in the world.

Wallace's education and that of his age and generation, taught him that the earth was solid and firm. These essential characteristics were self-evident in every hill and dale of his native England. Rocks may contain water, but never fire; except for its awesome reality, Wallace mused, the concept of a volcano, when first presented to a European, would not have been believed.

'Whence comes that inexhaustible fire whose dense and sulphureous smoke for ever issues from this bare and desolate peak?' he questioned. 'Whence the mighty forces that produced that peak and still from time to time exhibit themselves in the earthquakes that always occur in the vicinity of volcanic vents?'

Wallace would have enjoyed the half-joke, called Smith's Law, of present-day geologists, which rather whimsically states that anything that did happen, can happen.

Earthquakes and volcanos occur on the creaking and leaking margins of the Earth's 12 major tectonic plates. These are moving in various directions relative to one another, usually at speeds of around six centimetres a year. At the margins or seams, between the plates, three basic types of collision occur: compressional, where the plates override one another; extensional, where the plates break apart; and side-slipping or strike-slipping where the plates are sliding against one another. Volcanos usually form in linear belts because they mark the boundary of the first two types of interactions (see Figure 1, page 74).

At the compressional margins, or subduction zones, such as along the Sunda arc, one plate, usually oceanic, rides underneath the other. Friction causes earthquakes with epicentres from near the surface to depths of 700 kilometres, as one plate is shafted deep into the Earth. The ocean floor is warped downward where the plate plunges beneath the surface, forming a deep ocean trench. The edge of the upper plate is crumpled into folds or shattered into thrust faults. Rock layers are scraped off the plunging plate, just like butter scraped off a knife, to be smeared against the upper plate. The geological result is an island arc like those that make up the Indonesian Archipelago. Or a mountain chain like the Himalayas.

The plate shafting into the earth plunges at an angle, so that volcanism associated with subduction is usually about 100 to 200 kilometres landward of the deep ocean trenches. For instance, the belt of volcanos along the Sunda arc occurs only on the overlying Eurasian plate, directly above where the Indo–Australian plate plunges steeply beneath it (see Figure 1, page 74). The number of volcanos decreases gradually with distance away from the volcanic front towards Asia.

As the plate slides into the earth it drags with it water, carbon dioxide and minerals containing sodium and potassium—such as occur in oceanic sediments. These lower the melting point of magma and increase its gas content. Molten magma deep within the earth is like carbonated beverage in a bottle. It carries gas held in solution by the surrounding pressure. When that pressure is suddenly reduced, as in a volcanic eruption, the gas, like bubbles in champagne (but not nearly as pleasant), explodes out of solution. When this confined gas reaches the surface, via a volcanic vent, it is released explosively as volcanic ash, pumice, cinders, blocks and molten lava bombs. Together these are known as pyroclastics or fire fragments.

Expanding gas shatters magma into sand, grit and rice-sized fragments—collectively called ash—at the moment of eruption. Ash and water combine to form mud which avalanches down the sides of volcanos or even rains from the sky. Bubbly looking pumice forms when the rock is cooled so quickly that it traps the

74 WHERE WORLDS COLLIDE

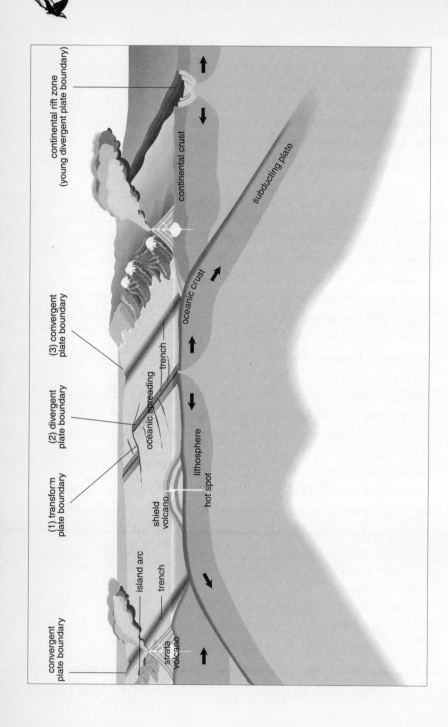

volcanic gases in tiny bubbles. Pumice, like glassy froth, is as light as styrofoam and floats on water. Volcanic blocks are usually fragments of old crater walls that are ripped loose in explosive eruptions. Molten lava bombs are clots of liquid lava which are flung into the air. As molten lava is quenched by the sea, some of it shatters into sand-sized particles. This new sand is swept along by shoreline currents and forms black sand beaches.

The different types of pyroclastics and the order in which they are produced give volcanos their characteristic shape. Alternations of volcanic ash and lava flows build beautiful, steep, nearly perfect cones with concave slopes, called 'strato-volcanos', like the one that Wallace had been looking at.

Because they are above the sea and easy to detect, 85 per cent of all volcanos known (400 of the more than 539), are these land-lubber volcanos. Surprisingly, these volcanos produce less than one-quarter of the world's magma. This is because the bulk of the planet's magma is produced by the other type of collision volcanos, the extensional, or rift, volcanos; the ones we cannot see.

Rift volcanos mark the spreading apart, or rifting, of major plates. In contrast to the explosive nature of the compressional volcanos which eject volatile, gas-filled rock 'contaminated' with crustal rock, the rift volcanos are characterised by the relatively non-explosive outpouring of fluid lava. These volcanos are

---

Fig 1. Cross section of plate tectonics.

A cross-section of the Earth's surface shows that it is composed of leaking and creaking tectonic plates. At the margins or seams of the plates three basic types of collisions occur: (1) where the plates override one another (transform plate boundary); (2) where the plates break apart (divergent plate boundary); and (3) where the plates slide against one another (convergent plate boundary). (After an illustration in Decker and Decker, 1981.)

'quiet' because of the high pressure under which they erupt and because the material that is ejected is not so volatile.

Indeed, this is the material that the Earth's tectonic plates ride on; a layer of partially molten, largely non-volatile material, 100 to 200 kilometres beneath the Earth's surface. Fractures in the Earth's crust at the extensional plate margins allow the partially molten rock to ooze to the surface, forming volcanos and broad, submarine mountain ranges. These extraordinary features can be tens of thousands of kilometres long, thousands of kilometres wide and kilometres high: drain away the oceans and the greatest mountain system on Earth would be exposed.

Indonesia leads the world in many volcano statistics. At 76, it has the largest number of historically active volcanos. And over the last 10 000 years (prehistoric time) at least 132 volcanos have been active. Indonesia's total of 1171 dated eruptions is only narrowly exceeded by Japan's 1274. Indonesia has suffered the highest numbers of fatal eruptions and more than anywhere else, eruptions which have produced mudflows, tsunamis, giant upwelling mounds of lava, called domes, and pyroclastic flows. It is a land of global-scale fireworks. Perhaps the most famous of these was the Krakatau eruption of 1883—21 years after Wallace had left the Indonesian Archipelago and one year after he acted as pall-bearer at the grave of his greatest friend, Charles Darwin.

To put the era into perspective, in 1883 the concepts of evolution, the germ theory of disease and heredity were in their first generation (words such as 'chromosome', 'electron' and 'radioactivity' were not yet known). The first gasoline-powered automobile was being trialled. This was the year that, thanks to brand new global telegraphic communication, news of a volcanic eruption was brought, for the first time, into people's homes around the world. Krakatau was destined for fame.

Situated in the middle of the Sunda Strait, between Sumatra and Java, Krakatau is located on one of the most historically

FIRE-SPITTING MOUNTAINS

travelled shipping lanes of the world, between China and Europe. The 1883 eruption had many witnesses. And the accounts are morbidly compelling.

Ships at sea were pelted by a hail of pumice, their decks accumulating metres of tephra (solid material). Crew members had holes burned in their clothes. Every breath was full of a hot, choking, sulphurous wind. Winds howled to hurricane force. The sky gloomed to an intense blackness—darker than the darkest night—punctuated by flashes of hell-fire. The pumice-rain turned to mud which fell from the sky on nearby villages, in sticky torrents. People more than 50 kilometres away on southeastern Sumatra sought shelter from a cloud of hot material which surged over the sea and left their skin peeling off them in strips.[58]

But while the pyroclastics were terrific, it was the tsunamis that caused the killing. Tsunamis are long-period sea waves that travel at speeds of up to 800 kilometres per hour and build to a devastating height on reaching land. They form when huge masses of water are displaced. Krakatau went into paroxysm during the early hours of 27 August. People as far away as Daly Waters in the Northern Territory of Australia heard the detonations that finally threw 18 cubic kilometres of the crater wall into the sea.

Shortly afterwards, the towns and villages along the shores of the Sunda Strait, already enveloped in a murky darkness, were overwhelmed by a succession of great sea waves. Rushing winds were driven before these tsunamis as they came on like immense dark walls. During the 1883 Krakatau explosions, waves reached 40 metres in height. They swept the kampongs along the nearby coastlines of Java and Sumatra into the sea. When they crashed onto the villages, not even rooftops were visible. Tsunamis caused the loss of 36 380 lives that night.[58] Two hundred and twenty kilometres from the area, carcasses of wildlife, including tigers, were spotted among the 150 or so human corpses found at sea. Enormous tree trunks were borne along by the current.

Coral blocks up to 100 tonnes in weight were thrown onto the coast 50 kilometres away. The largest wave, following the largest explosion, carried a steamship inland two-and-a-half kilometres (killing the crew of 28).[58] Heard 4653 kilometres away on Rodriguez Island, this explosion hurled ash into the stratosphere. Jakarta was darkened by ash. Ash fell on the Cocos Islands, about 500 kilometres to the south, on the same day. When sunlight finally threw light on Krakatau it revealed an island that had been disembowled. Where the middle of the island had been, a circular submarine depression, a caldera, six kilometres in diameter, had formed.

The next day, sea level drops near Bombay, 4500 kilometres from Krakatau, were so sudden that the fish were left floundering on the mudflats. Waves from the explosion travelled the world, reaching France 32 hours later and 17 960 kilometres away. Fine ash fell on Scotia 5321 to the north-west, 10 days after the paroxysm.

Captain A.W. Newell, of the Boston barque *Amy Turner*, brought in some pumice which was washed aboard his vessel on 17 September 1883, about 271 kilometres south-west of Krakatau. It covered the sea in windrows. In November, pumice was reported 2800 kilometres west of Krakatau. Sailors were said to have walked about on patches of it.[58] Pumice floated on the open ocean for a long time. More than a year after Krakatau went into paroxysm, pumice was washed ashore at Durban, South Africa, 8170 kilometres away. In July 1884, pumice and skeletons landed on Zanzibar beaches, 6170 kilometres west of Krakatau. Sailors continued to report the novelty of floating pumice well into 1885.[58]

The dust that was lifted into the stratosphere circled the globe, producing astonishing visual effects: dramatic sunrises and sunsets and a blue-green appearance to the sun and moon. The dust spread westwards, encircling the equator in two weeks, then drifted both north and south. The sun arose green in the South China Sea, 12 days after Krakatau's paroxysm. In Madras, India, 3500 kilometres away, the sunsets were reported to be a bright pea-green and the moon, near the horizon, was pale green. Three months later, in November, spectacular sunsets were observed in London and around the United States where, in New York City, fire engines were summoned one November night because of the brilliant, fiery sunset afterglows.[58] Over the next three years, average solar radiation in Europe decreased 10 per cent and average world temperatures were less than normal.

The 1883 eruption of Krakatau was not its first. Ancient Javanese scripts chronicle an eruption of Krakatau in AD 416: 'A great, glaring fire, which reached to the sky, came out of the Krakatau. The whole world was greatly shaken. Violent thundering was accompanied by heavy rains and storms. The noise was fearful. At last the mountain released a tremendous roar and burst into pieces and sunk into the depth of the earth. The sea rose to inundate the land.'

The chronicles continue that the country to the east, now the western tip of Java and the country to the west, now the eastern tip of Sumatra, was inundated by the sea. The inhabitants were drowned and swept away with all their property. After the water subsided, the mountain and the surrounding land became sea and the island of Java was divided into two parts. This, according to the chronicles, is the origin of the separation of Sumatra and Java.

That volcanos can easily produce large chunks of land and just as easily tear them apart is evidenced by the eruption of Kilauea volcano which has increased the island of Hawaii by 4000 square kilometres over the past several years.

While the 1883 Krakatau eruption became perhaps the most

famous of the world's volcanic eruptions, it was not by any means the biggest. Krakatau disgorged 18 cubic kilometres of pumice and ash. History's largest eruption was on Sumbawa—east of Bali, between Lombok and Flores. Here, in 1815, Tambora disgorged 150 cubic kilometres of rock, lava and ash; eight times the amount of material from Krakatau. Tambora shook Sumbawa with a gigantic explosive eruption which left the 4000-metre high mountain 1250 metres lower and with a crater 11 kilometres across. Some 92 000 people died as a result of the eruptions and a region extending up to 600 kilometres west of the volcano was plunged into darkness. At the time, Tambora was thought to be inactive. Recent evidence estimates that 40 cubic kilometres of dust and ash found its way into the upper atmosphere, causing global cooling, summer snowstorms and crop failures on the other side of the planet. In fact, 1816 was remembered in Europe and America as the year without a summer.[58]

As mind-boggling as this seems, the Tambora eruption would have been dwarfed by Toba in north-western Sumatra. Data on this prehistoric eruption, about 75 000 years ago, indicates that Toba let loose 2800 cubic kilometres of material, the likes of which we have never seen in history. The eruption created the 100-kilometre-wide Lake Toba, the largest caldera in the world.

The reverberations of this 75 000-year-old event still have manifestations in today's wildlife: 17 of the region's bird species do not extend south of Lake Toba and 10 do not extend to its north. Several mammals also have split distributions, severed by Lake Toba. These include Thomas' Leafmonkey, the White-handed Gibbon, Orang-utan, Tarsier, Banded Leafmonkey, Sumatran Rabbit and Tapir. It is hard to imagine the devastation caused by such an explosion and the time taken to heal it.

Krakatau, however, provides at least a hint of how such an

area of essentially sterile earth can be recolonised. The 1883 eruption—the equivalent of 2000 Hiroshima bombs—resulted in Krakatau being stripped entirely of life. But in 1886, 26 plant species had already colonised Krakatau; in 1897, 61 species; and in 1906, 108 species were collected. From 1908 to 1928, 276 species were collected and at least six different plant communities were in evidence. Succession seemed to be actively taking place, but in 1979 the number of species were found not to have increased.[58]

A closer look, however, revealed that while the number of species stabilised, the composition of those species had changed. It appears that while two to three species a year may be dispersing to the Krakatau complex, the same number are going extinct due to random events. In the century or so since the Krakatau eruption, all except one of the species have been pioneer species, having their seed distributed by wind and ocean current. In time, they will build an environment for the more permanent, mature phase of forest development.

It has taken a century for one tree typical of a primary forest to take hold on Krakatau. Unfortunately, at the present rate of forest destruction on Java and Sumatra, it will take much less than a century to drain dry the pool of parent forest trees. At this rate Anak Krakatau—the child of Krakatau—will never see the primary forests of its great parent.

Volcanos 'live' for roughly one million years, though they can erupt for up to 10 million years. In the end, erosion takes over and the stumps of vanquished volcanos join the other passive mountains of the world. Oceanic volcanic islands like Hawaii, however, are washed completely away by the sea in five to 10 million years and then slowly subside beneath a growing coral atoll cap.

Living along the Nusa Tenggara chain of islands has meant three million years of living dangerously. And while such volcanic action must surely have filtered the plants and animals that could disperse along the chain of islands between Lombok and Timor, it does not really explain why, for instance, the avifauna of Lombok is so strikingly different from that of Bali. After all, both islands have active volcanos. What is it that makes this section of the Wallace Line so potent?

CHAPTER 5

STEGOLAND

CHARLES DARWIN provided Wallace with a hint. Darwin wrote to Wallace from London pointing out that, elsewhere in the world, he had found a relationship between sea depth and known patterns of mammalian distribution. The deeper the sea, the more distantly related the fauna; the shallower the sea the more closely related the fauna. The ever-insightful Darwin had discovered that depth was a measure of time.

Shallow seas implied a recent connection. In England, for instance, with only a few exceptions, nearly every mammal, bird, reptile, insect and plant is found on the adjacent continent. However in Sri Lanka, closer to India than Britain is to Europe—but with a deeper sea in between—many animals and plants are different and peculiar to the island. In the Galapagos near South America and Madagascar near Africa, almost every living thing is endemic. Both are separated from their nearest continent by deep sea trenches.

This fitted with Wallace's own observations that the (to this day) little-known Aru islands, in the Arafura Sea north of Australia, and the islands of Misool and Waigeo to the south and west of the Vogelkop Peninsula respectively 'agree with New Guinea in their species of mammalia and birds much more closely than they do with the Moluccas and we find that they are all united to New Guinea by a shallow sea'.

Borneo, Sumatra and Java clearly lay on the same shallow bank or continental shelf. Indeed, so shallow was the sea that ships could anchor anywhere in the vast seas between the islands. That the faunas of these well-known islands resembled each other was obvious. In fact, Wallace thought the faunas were as similar as distant areas could be expected to be, even without an intervening sea. The sea must have been a recent invader, flooding the lowlands and deep valleys and isolating the highlands and volcanic peaks that became Borneo, Java and Sumatra. The Wallace Line more or less marks the edge of the Asian continental shelf.

Wallace had no explanation for the changing sea levels. In his day it was adventurous enough just to suggest that the surface of the earth changes over geologic time and downright risqué to point out that, as a result, the forms of life that inhabit that surface have also slowly changed. Nevertheless, he could not resist having a stab at an explanation. Perhaps, he thought, the great volcanic chain along the islands of Sumatra and Java had something to do with it since they were constantly spewing out what must have been the very foundations of the land. Wallace imagined that the land, emptied of these foundations, would slowly collapse and slump beneath a steady-state sea.

The reality is much more bizarre! And as imaginative as Wallace was, naturalists of the day would really have thought him to be in 'Wonderland' had he suggested, from his womb-warm environment, that he lived in an ice age!

Nothing could be further from the truth than to consider today's climate normal. The last time conditions were like they are today was 125 000 years ago, and then only for a short period lasting maybe 4000 years, during an interglacial. The present ice age is characterised by regular glacials, punctuated by these short interglacials (one of which we are currently in). We are, ourselves, ice age animals, having evolved during the past 2.5 million years or so, during which there have been 25 recognisable glacial–interglacial alternations, following a curious 100 000 year cycle. During these cycles the great continental icesheets of the northern hemisphere and Antarctica wax and wane, in turn locking-up and releasing water. Far from being steady-state, the sea levels rise and fall, from levels similar to those of today to levels as low as 230 metres below present levels.[37]

This dizzying Pleistocene climatic yo-yo was initiated with the union of South America and North America, around 3 million years ago, across the umbilical-like Panama Isthmus. Before the union, the waters of the Pacific and those of the North Atlantic mixed freely, stopping any strong currents forming. This allowed warm tropical waters to filter poleward and cold polar waters to move equatorward. The union of North and South America, however, locked-out the levelling influence of the Pacific and allowed the North Atlantic Gulf Stream to strengthen into a strong, circular flow which acted as a barrier to any flow of warm water to the Poles, thus locking-in and strengthening the cold. The same sort of thing had happened earlier to Antarctica when Australia surfed northward, allowing the bitingly cold circumpolar current to strengthen and engulf Antarctica. With Antarctica already sheathed in ice, the stage was now set for the expansion of the North American icesheets and the birth of a full-blown ice age.

It was the cosmic flutter of a butterfly's wings that tipped the balance. The Earth orbits the sun in an ellipse. Every 100 000 years there are slight changes in this ellipticity and it is these almost imperceptible changes that ramify, on Earth, into the 100 000 year glacial–interglacial–glacial cycles. At their glacial maximum, the mammoth icesheets have a major influence on world climates and world sea levels.

At these times, Borneo, Sumatra, Java and its offshore island Bali, and the Malay Peninsula are connected by dry land. The Lesser Sundas, or Nusa Tenggara, comprising Lombok, Sumbawa, Komodo, Flores and the Alor islands emerge as a thin peninsula separated from Bali by a very narrow channel. New Guinea, Australia and Tasmania form a single land mass, called Meganesia. This land mass also includes the offshore islands of Misool, Waigeo and the Aru group. This vast land is separated from Timor by as little as 70 kilometres. Timor itself is not much more than a stone's throw from Alor island. A string of islands probably provides a tenuous link between Sulawesi and part of Nusa Tenggara. And Borneo is connected through a thin leash of land, now Palawan, to the Philippines.

During glacials, rivers gouged out deep valleys which today appear as submarine channels on the sea floor. A huge North

---

Map 18. Palaeogeographical reconstruction of the Sunda–Sahul region during Pleistocene times.

Shown here are the courses of major rivers during the Pleistocene glacials, around 12 000 years ago. Today they are submarine channels on the sea floor. (After Kottelat et al., 1993 and Kitchener et al., 1990.)

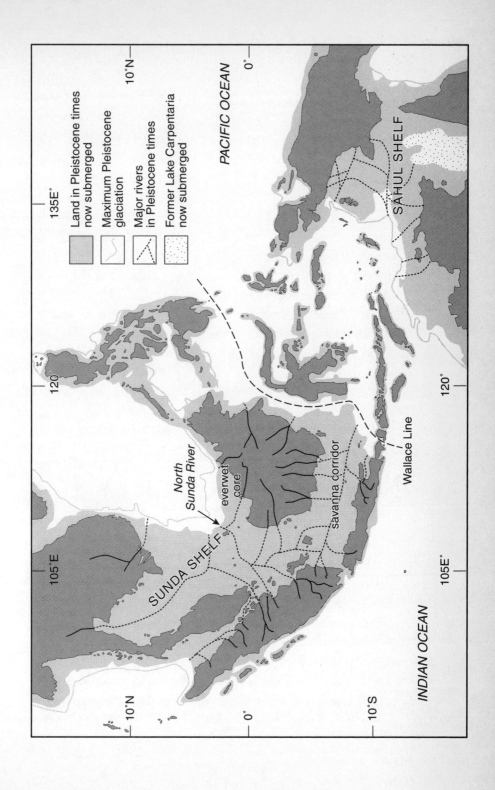

Sunda River flowed northward to the present South China Sea. It collected its tributaries from eastern Sumatra, western Borneo and from Malaysia and other parts of the east coast of mainland Asia. Another river flowed eastwards, throwing its fine sediment into the sea north of Bali. It embraced a catchment extending from south-eastern Sumatra, northern Java and southern Borneo. On Meganesia, two separate rivers flowed westwards. One from southern New Guinea entered the sea north of today's Aru islands and the other drained an area which included today's Gulf of Carpentaria, Top End of the Northern Territory and Kimberley Region.

Interglacials, characterised by sea levels similar to today's, have all been brief events lasting perhaps 5000 years, with a saw-toothed build-up towards full glacial conditions. If anything can be characterised as normal in the Pleistocene, the period in which *Homo sapiens* evolved, it is a climate tending toward glacial.

What sort of impact did these drastic changes have on the land? Ancient pollen provides a clue. Undisturbed by earth movements or the activity of humans, an extinct crater on the Atherton Tablelands in northern Queensland, Australia, has been steadily filling with sediment for over 125 000 years. The sediment, containing pollen from the forests of the surrounding catchment, has been swept down and layered in a neat time-series. A drill-core from Lynch's Crater provides one of the few windows to check on modern interpretations of the region's past.[80] Using the dustings of preserved pollen, a fingerprint of the landscape during the last interglacial–glacial–interglacial cycle appears.

Before 50 000 years ago, a rich tropical rainforest flourished on the volcanic soils of the Atherton Tablelands. Around 50 000 years ago this rich forest withered, under drying conditions, to become a cooler conifer-dominated rainforest. As the region entered the glacial peak of 18 000 years ago, rainforests were entirely obliterated. They were replaced by a sclerophyll forest dominated by *Eucalyptus* and *Casuarina*. It would have needed a rainfall plunging to 20 per cent of its former level to have caused this change. Only 10 000 years ago, as the climate ameliorated into that of the current interglacial, did the rich flowering forests return. And this was only one of 25 similar cycles that have occurred in the last 2.5 million years!

The scanty pollen evidence from Asia, from a core extracted from peat sediment near Kuala Lumpur, Malaysia, also reveals pollen assemblages dominated by coniferous *Pinus* and grasses. *Pinus* does not occur in present-day Malaysia. The combination is suggestive of the pine woodlands of Thailand which grow today only under a markedly seasonal environment. No rainforest pollen was found in this core. Pollen records from Brunei and from much of the South China Sea, which forms an ice age plain, also suggests a seasonal climate, like that of northern Australia and southern India where savanna is one of the dominant vegetation types.

For these sorts of changes to have occurred, temperatures would be perhaps 3.5 degrees Celsius cooler and rainfall half and in places, only 20 per cent that of today. Falling sea levels and a greater land area help to explain these changes, as do the differing configuration of land and water. For instance, ocean currents which today enter the Malay Archipelago through the Torres Strait, the South China Sea and south of Mindanao in the Philippines would have been severed. Their buffering effect, on a climate already sucked dry by the polar icesheets, was removed. The climate would have become virtually continental.

What would such an environment feel like to evolving *Homo*? Proto-humans would be familiar with glaciers on the highest mountains of Borneo and Sumatra as well as an expanded version of those on the island of New Guinea. In the cooler, drier climate, montane forests extended to the flanks of the hills. The rainforests shrank to moist refuges, closer to the equator on New Guinea, Borneo and Sumatra.

Except for these areas, which lay on the coasts of the great continents of Meganesia and Eurasia respectively, rain fell in a distinct, short season. Seasonal forests, such as monsoon forests and grassy savanna scattered with shrubs, trees and palms, thrived. In Asia, these trees and shrubs were selected from the retreating rainforests for their toughness and resistance to fire and drought. Savanna, intermingled with monsoon forest and rainforest nestled in moist valleys or along rivers, forming a patchwork corridor from mainland Asia to New Guinea. *Eucalypt*-dominated savanna woodlands grew on the infertile lateritic plains from New Guinea to Australia around a vast, shimmering, salty lake in the vicinity of the Gulf of Carpentaria. Here, marsupial rhinos, rhino-sized marsupial diprotodons, giant kangaroos, great flightless birds and 200-kilogram, horned turtles roamed. They would have been prey to the rapacious giant goanna, perhaps four times the size of the killer Komodo Dragon. These more open conditions, easy to move in and easy to burn, were highly favourable to large browsing and grazing animals and to hunter–gatherer groups of people.

Early Pleistocene proto-humans would have arrived at the edge of Asia to hunt the last of a relictual population of *Mastodon*, which became extinct in tropical Asia around 2 million years ago. Mastodontids arose in Africa around 30 million years ago, giving rise to elephants and the now-extinct stegodonts. The last of the line of mastodonts evolved into the famous glacial-loving Woolly Mammoth *Mammuthus primigenus* which roamed the Siberian steppes in the late Pleistocene. This animal went extinct only around 8000 years ago. Mastodontids, however, were not generally animals of the icy steppe but, rather, animals of the woodlands—just like the forests and woodlands of an expanded, cooler and drier continental Asia, which included the savannas of Sumatra and Java.

That most of the Pleistocene was indeed characterised by these savanna environments is confirmed by the sequential palaeontological fossil record of Javan mammals.[65] Though incomplete, the Javan mammalian biostratigraphy reveals a fascinating faunal succession through seven different ages. More akin to scenes of Africa than tropical Asia, over the past two million years, stegodon, ancestral hippopotamus, antelope and deer were abundant, as well as large carnivorous cats and ancient canines. Later, tapirs, hyaena, rhinoceros, modern elephants and large cattle-like bovids arrived and, with the stegodons and pigs, evolved into different forms.

The nature and size of these animals indicate that for most of the Pleistocene, Java was a wedge of land at the edge of a continent. Herds of grazing and browsing mammals roamed large tracts of open forests. These were prey for tigers, now-extinct cats, ancient canines and hyaenas; a carnivorous complement characteristic of continents. Indeed, mammalian carnivores are the first animals to become extinct under island conditions, as they rely on the abundant wildlife, usually found on continents, to survive.

Of the seven or so different suites of fossil mammals, only one group that represents the fauna of an interglacial can be identified. Around 125 000 years ago, tropical Java was characterised by a high quantity of rainforest-loving Orang-utan. Alongside the Orang-utans was a species of pig, which is still extant and a typical animal of the rainforest floor where it roots out much of its food from the moist soil. The modern elephant species *Elephas maximus* also makes its first appearance in the fossil record here. Today this Asiatic Elephant is also a denizen of the denser forests.

At the height of the Pleistocene glaciations, land mammals,

including proto-humans, made their way from South-East Asia to what are now the islands of Sumatra, Java and Borneo. The Makassar Strait, the northern end of the Wallace Line between Borneo and Sulawesi, however, remained a wide and deep barrier to the dispersal of most mammalian species of Asiatic origin. The Orang-utan, tiger, panther, Malay Bear, elephant, Malay Tapir, Giant Pangolin, Javan Rhinoceros, Sumatran Rhinoceros and Banteng Cow have never been found in Sulawesi, either living or fossil. Even during the Pleistocene glaciations, the northern end of the Wallace Line has maintained its integrity.

The Nusa Tenggara chain of islands extending from Bali to Timor, however, was accessible—at least to great travellers like elephants, stegodons, giant turtles, Asian crocodiles and giant pigs. These animals not only managed to hop the narrow water gap from Bali to Lombok, they also managed to vault from Java and Flores to the south-western arm of Sulawesi and from the Alor islands to Timor during low sea levels. Certainly, the two modern species of elephants are very strong swimmers capable of swimming 45 kilometres under extreme circumstances and strongly inclined to swim shorter distances of several kilometres in search of food.[32] It seems they inherited this from their ancestors.

Scattered bones in the river terrace deposits in the south-western arm of Sulawesi are all that remain of an amazing saga. Here, in the early Pleistocene, around two million years ago, some ancient elephants, which had languished in poor condition on a seasonally drought-stricken peninsula, had sensed the unplundered resources of a new land. Across a shallow and narrow arm of the sea, they swam the gap between Nusa Tenggara and Sulawesi. Later, the sea rose and the gap became an ocean. As the ocean continued to rise—up to 25 metres above the present level—the sea flooded a low valley and the south-western arm of Sulawesi was cut off from the rest of Sulawesi.[78] Stranded on an island, the elephants dwarfed and evolved into the pygmy *Elephas celebensis*, virtually identical to its direct ancestor, but half its size. From Java, *Stegodon sompoensis* had also made it across to Sulawesi. It too underwent a 50 per cent size reduction, evolving into the pygmy *Stegodon trigonocephalus*.

In Flores, the discovery of another pygmy stegodon, closely related to *Stegodon trigonocephalus*, indicates that stegodons were sauntering along the thin peninsula which formed when Nusa Tenggara emerged, jutting out from the edge of Asia, during the times of glacially lowered sea levels. During these times they must also have swum the small gap across to Timor. In all of the three islands—the south-western arm of Sulawesi, Flores and Timor—there is evidence of continental-sized stegodonts besides pygmy forms. This indicates that the larger animal gave rise to the insular dwarfed forms during the times of raised sea levels when the sea gaps became too wide to swim. Eastern Indonesia must have been a veritable 'Stegoland', an incredible ice-age refuge for the pygmy stegodons.[28] Fossil records of Pleistocene fauna for the rest of the archipelago, however, are rarer than stegodons' teeth.

'Plants have much greater facilities for passing across arms of the sea than animals. The lighter seeds are easily carried by the winds and many of them are specially adapted to be so carried. Others can float a long time unhurt in the water and are drifted by winds and currents to different shores ...'

In reality, plants are more revealing of ice-age conditions. A shame that Wallace was not in the least bit interested in them from a biogeographic perspective, apart from some obvious exceptions such as the fragmented distribution of the flora of mountain tops (the subject of Chapter seven) and rather token comments like the one above. He did not mind admitting it, putting it down to the fact that the flora was imperfectly known. In many respects Wallace found plants a nuisance, since they got in the way of collecting animals, a sentiment not unfamiliar among zoologists today. Plants, in turn, pay scant attention to zoologists' biogeographic boundaries.

Boundaries, or rather breaks in distribution, do, however, exist for plants. One particularly significant type of fragmented distribution is the disjunct distribution of savanna and monsoon plants between continental Asia and eastern Indonesia.[77] This disjunction is characterised by savanna and monsoon species of plants which occur in both Indochina and Australasia but not at all in between—in Borneo, Sumatra and Malaysia. The kinds of historical change which would cause such a distribution are sea-level changes, changes in the degree of seasonality in rainfall and temperature changes: that is, an ice age. A reduction of sea level would provide land linkages from mainland South-East Asia and Australia out to the islands. An increase in seasonality would make these linkages available to plants requiring a seasonal climate thus assisting migration between, for instance, Thailand and northern Australia.

Today, populations of these monsoon-climate plants are separated from each other by sometimes 2000 kilometres. Yet remarkably there are very few differences between the isolated populations, indicating a continuous range across South-East Asia, Malaysia, Sumatra, Borneo and Java, extending into New Guinea, as well as into the savannas of Australia in the very recent past.

Clearly, the vicissitudes of the Pleistocene climate have had a major impact on the evolution of the flora and fauna of the Malay Archipelago. The spasmodic expansion and contraction of environments has driven dispersal and speciation. On a background of shifting continents, it is these climatic fluctuations that have largely created the distribution of plants and animals of today's Indonesia. And it is to these present-day distributions that we must turn to help unravel ancient tracks.

On present-day distributions, the Wallace Line between Bali and Lombok, separates adjacent faunas that are quite different. Unlike the robust biogeographic barrier of the Makassar Strait which isolates distantly related fauna, the barriers between Java–Bali, and Lombok and the rest of Nusa Tenggara, merely separate a rich continental fauna from an island fauna. Even at the height of glacials, this island fauna was limited by the size of the islands.

CHAPTER 6

ISLANDS IN THE SEA

B ACAN, 1858. Wallace leaned back on his comfortable rattan chair and picked up a scientific paper from among the scattering of books, penknives, scissors, pliers, pins and bird labels that lay on the rude table. He had received the paper in a batch of literature which arrived in Ternate by mail-steamer. Ternate, the fourth of a row of fine volcanic islands which skirt the western coast of Halmahera, is a major administrative centre of the Moluccas (today's Maluku). One of the famed Spice Islands, it was formerly worth its weight in gold. Ternate became Wallace's base for three years—a place where he could pack his collections, recruit his health and collect his mail before heading off into unexplored territory; like the island of Bacan, south of Ternate.

It was bucketing rain outside but the palm-leaf thatch roof of Wallace's temporary abode proved remarkably waterproof. The road in front of the tiny hut was a river of mud, leading through cultivated fields to a forest about one kilometre away. Wallace sighed. Birds and insects would be rare indeed in this weather.

He glanced at his freshly killed, vaguely noisome collection, arrayed for drying on shelves suspended from the bamboo struts of the roof and lined with ant-proof, insulating oil cups. A line stretched across one corner of the room carried his freshly rinsed cotton clothing and on a bamboo shelf was arranged his small stock of crockery—teapots, tea cups, teaspoons—more reminiscent of the Mad Hatter's tea party than of a (slightly mad) naturalist in the wilderness. In an adjoining annex, two bamboo chairs and a roughly made bamboo bedstead surrounded by mosquito netting and some quaint Scottish plaid for privacy, completed his furnishings. Boxes containing his collecting gear and prepared specimens were neatly stacked on the earthen floors against the thatched walls. In the cramped space, he would have had to be careful unfolding his rather long, wiry, body. He would not, at any rate, have been able to stand straight since the ceiling was low. Perhaps it was this that made him habitually walk with a slight stoop.

Wallace had stayed in worse conditions. In Lombok he had only one small 'parlor' room in the hut which he shared with a family. This room, he recalled, 'had to serve for eating, sleeping and working, for storehouse and dissecting room; in it were no shelves, cupboards, chairs or tables; ants swarmed in every part of it and dogs, cats and fowls entered it at pleasure'. Wallace's principal piece of furniture at this time was a box, which served him as a dining table, a seat while skinning birds and as the receptacle of the birds, when skinned and dried. To keep them free from ants he borrowed an old bench, the four legs of which were placed in coconut shells filled with water. The box and the bench were the only places where anything could be put away, 'and they were generally well occupied by two insect-boxes and about a hundred birds' skins in the process of drying'. 'All animal substances', Wallace wrote matter-of-factly, 'require some time to dry and emit a very disagreeable odour while doing so, which is particularly attractive to ants, flies, dogs, rats, cats and other vermin.' As if this were not enough, the room still actively functioned as a parlor for his host and Wallace found himself apologetically working around his host's guests who no doubt visited more frequently than usual to be entertained by this most bizarre spectacle.

Wallace thought of Lombok as he read the recently published scientific paper by Philip Lutley Sclater.[56] In this paper, Sclater proposed that the world could be divided into six faunal regions based on the distribution of birds. Being deeply religious, Sclater had said that these represented six distinct centres of creation. No doubt ignoring that, Wallace saw that the regionalisations, nevertheless, had merit because they corresponded to more generalised biogeographical patterns of distribution, based on his evolving ideas of evolution. Wallace, however, took issue with Sclater's boundaries. In particular, Sclater had placed the Malaysian Peninsula, Sumatra, Java, Bali, as well as the Lesser Sundas (today's Nusa Tenggara)—which Wallace considered was influenced by Australia—in the Oriental region.

Wallace scratched a note to *The Ibis*, a learned journal of ornithology.[71] Of the boundary between the Oriental and the Australian region Wallace wrote (amongst other things), '… its south-eastern limits I draw between the islands of Bali and Lombok. ... Barbets reach Bali but not Lombok; *Cacatua* (parrots) and *Tropidorhynchus* [now *Philemon* (friar-birds)] reach Lombok, but not Bali: this I think settles that point'.

While the Wallace Line was officially accepted as the boundary between the Oriental and Australian region, the point remained far from settled.

Actually, the influence of Australia is felt well to the west of Lombok. Wallace, too, noticed that the south-east tradewinds, lasting about two-thirds of the year (from March to November), blow over the Australian continent, 'producing a degree of heat and dryness which assimilates the vegetation and physical aspect of the adjacent islands to its own'. The enormous rainshadow of Australia, extending from Timor to eastern Java, turns the 'damp, gloomy ever-verdant forests' into 'deciduous, parched-up forests of prickles'.

In eastern Java these drier forests already form a barrier, stopping some characteristic west Javan species extending eastward.[76] Wallace, in western Java, obtained a brightly coloured iridescent yellow and green trogon *Harpactes reinwardtii*, a gorgeous little minivet flycatcher *Pericrocotus miniatus*, 'which looks like a flame of fire as it flutters among the bushes' and the 'rare and curious black and crimson oriole [*Oriolus cruentus*]', all birds not found in eastern Java.

In eastern Java he noticed another, more perplexing distribution pattern. Several species of birds were found in Java, Thailand and India, but not in Sumatra, Borneo or Malaya

Map 19. Palaeogeographical reconstruction of the islands of Bali and Lombok during Pleistocene times.

A southern corridor of dry land is likely to have linked Bali to Lombok via Nusa Penida during the Pleistocene, around 12 000 years ago. It is believed that Lombok was joined to Sumbawa for long periods during these glacial times. (After Kitchener et al., 1990.)

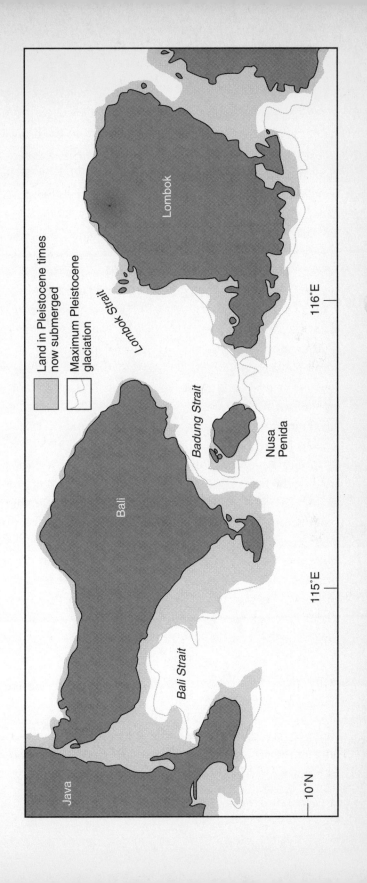

which were in between. The only way Wallace could account for a distribution 'so strange and contradictory' was by assuming that at some time in the not-too-distant past (considering the close relationship of the birds), Borneo, Sumatra and Malaysia, sank willy-nilly beneath the sea leaving Java high and dry as a receptacle for South-East Asian birds.

In fact, the 30 or so birds, being largely birds of the drier forests, share the same disjunct distribution pattern of savanna and monsoon plants. Echoes of the recent past, birds such as the Green Peacock, Brown Prinia and Common Tailorbird[42] flourished across South-East Asia when Java and Bali formed part of a drier monsoon and savanna-covered mainland. Today, only narrowly separated from Java by a shallow strait, fully 97 per cent of Bali's 172 birds are shared with Java, a fact not missed by Wallace.

'During the few days which I staid on the north coast of Bali, on my way to Lombok, I saw several birds highly characteristic of Javan ornithology. Among these were the yellow-headed weaver, the black grasshopper-thrush, the rosy barbet, the Malay oriole, the Java ground-starling and the Javanese three-toed woodpecker. On crossing over to Lombok ... I never saw one of them, but found a totally different set of species, most of which were utterly unknown not only in Java, but also in Borneo, Sumatra and Malacca. For example, among the commonest birds in Lombok were white cockatoos and three species of honeysuckers, belonging to family groups which are entirely absent to the west.'

These essentially Australian birds were thriving in Lombok's 'parched up forests of prickles'. These forests were more than a thorn in Wallace's side. 'The bushes were thorny, the creepers were thorny, the bamboos even were thorny. Everything grew zigzag and jagged and in an inextricable tangle, so that to get through the bush with gun or net or even spectacles was generally not to be done.' It was, however, in such places that the gorgeous green and glittering blue pitta *Pitta elegans* often lurked—the bird Wallace most sought on Lombok.

Not one to be put off, Wallace worked out a strategy to obtain this shy bird of the forest floor. 'At intervals they utter a peculiar cry of two notes, which when once heard is easily recognised and they can also be heard hopping along among the dry leaves. My practice was, therefore, to walk cautiously along the narrow pathways with which the country abounded and, on detecting any sign of a Pitta's vicinity, to stand motionless and give a gentle whistle occasionally, imitating the notes as near as possible. After half an hour's waiting, I was often rewarded by seeing the pretty bird hopping along in the thicket. Then I would perhaps lose sight of it again, till, having my gun raised and ready for a shot, a second glimpse would enable me to secure my prize and admire its soft puffy plumage and lovely colours.' Distributed from Africa to Australia, the brilliantly coloured pittas are found in both Bali and Lombok.

Lombok, in fact, shares more than half its birds with Bali and Java. The larrikin Australian birds, which have a habit of drawing attention to themselves, are only a minor, albeit noisy, component of the avifauna. It seems that Wallace, on the edge of a major scientific discovery, could not resist carefully selecting the data to support his ground-breaking theory. The 'highly characteristic' Javan birds he selected were also carefully chosen to make his point. They are all either poor dispersers (unlike the pitta, which is also Oriental), or birds characteristic of tropical rainforests, rather than the drier, deciduous forest.

Nevertheless, if 50 per cent of the birds between Lombok and Bali are shared, 50 per cent of them are not. Compared with the 97 per cent of birds that Bali shares with Java, the change between birds shared between Lombok and Bali is indeed noteworthy. This, more than the presence of a few Aussie adventure travellers, is the point of real biogeographical significance.

The significance of the Wallace Line between Bali and

Lombok is that it marks the end of a continental avifauna and the beginning of an oceanic one. Here, most families, genera and species of the rich continental Oriental avifauna are unable to cross the gap from the stable habitats of a continent to a changing constellation of oceanic islands.

Amazingly, the idea that there are indeed islands which have never been connected to the mainland was not an acceptable theory in Wallace's time. 'This' Wallace said, explaining the impeccable logic of Victorian England, 'would imply that their animal population was a matter of chance; it has been termed the "flotsam and jetsam" theory and it has been maintained that nature does not work by the "chapter of accidents".'

Wallace said that, on the contrary, 'we have the most positive evidence that such has been the mode of peopling the islands. Their productions are of that miscellaneous character which we should expect from such an origin and to suppose that they have been portions of Australia or of Java will introduce perfectly gratuitous difficulties and render it quite impossible to explain those curious relations which the best-known group of animals (the birds) have been shown to exhibit'.

In fact, in Maluku examples of haphazard colonisation abound. The Mountain Tailorbird *Orthotomus cuculatus*, for instance, is present in Seram, Buru and Bacan, but not Halmahera, Sula or Obi, while the Island Verditer Flycatcher *Eumyias panayensis* is present in Seram and Obi but not in Buru, Bacan, Halmahera or Sula.[8] The recent discovery of two new waterbirds for Seram: the comb-crested Jacana *Irediparra gallinacea* toward the western limit of its spread from Australia; and the White-breasted Waterhen *Amaurornis phoenicurus* on the eastern limit of its spread from South-East Asia, exemplifies the ongoing nature of colonisation.[8] Clearly the spread of species from west-to-east and east-to-west is a fluid process where small differences in the initial colonisation of an island can result in chaotically different faunas.

In Maluku, the outcome of this dynamic process in terms of endemic species is a little like a raffle: 'Thus Morty [Morotai] Island has a peculiar kingfisher, honeysucker and starling; Ternate has a ground-thrush (Pitta) and a fly-catcher; Banda has a pigeon, a shrike and a Pitta; Ke [Kai] has two fly-catchers, a Zosterops [white-eye], a shrike, a king-crow and a cuckoo; and remote Timor-Laut ... has a cockatoo and lory as its only known birds and both are of peculiar species.'

An island is nothing like a continent. Compared with a continent's stable patchwork of interlocking habitats, an island provides mere scraps of habitat, loosely connected and unstable and surrounded by expanses of sea water.

These act like a species' filter so that only animals which can survive an ocean crossing will make landfall into the forbidding island environments. Some large birds and bats can reach even distant islands. Smaller birds, bats and insects can be carried passively on high winds. Other animals such as rats make the journey on drifting rafts of vegetation. Rarely, animals such as pigs, deer and elephants will swim to an island, but they will normally need to sense from their continent or continental island what it is they are getting themselves onto. Certain groups of animals are completely unable to disperse over water barriers; even in birds and bats there are many species which have a total inability to disperse. Apart from notable examples such as bats and rats, mammals are very poor dispersers.

The origins of animals that actually manage to colonise oceanic islands are not always clear. Certainly the Galapagos has ultimately derived its birds from tropical America. Hawaii and the Malay Archipelago, however, have derived their birds from more than one continent. Very old islands become evolutionary centres in their own right, spawning their own species and—in time—even genera and families, such as Madagascar and Sulawesi. In all these circumstances, it is nonsense to artificially assign oceanic islands to one of the six continental faunal regions of the world defined by Sclater and Wallace. Ironically, it is precisely this regionalisation that, when applied to islands, produces the 'perfectly gratuitous difficulties' Wallace spoke of.

Nothing suits evolution better than a new, inherently unstable environment with a population of flotsam and jetsam individuals. Like a loose cannon, these individuals have an unconstrained selection of the genetic material of the mother population. In an evolutionary equivalent of a fast learning curve, only the fittest—those able to change rapidly—will survive. Inevitably, the evolution of island populations will be directed away from that of the mother population (although any flow from that mother population will tend to dilute this effect).

The result is that, while islands are impoverished in absolute numbers of species, they have an extraordinarily high rate of endemism. Of the total of 385 butterfly species recorded in Maluku[67] for instance, 82 are not known to occur elsewhere. Compared with the two per cent endemism for butterflies of the mainland Malay Peninsula, this 21 per cent endemism is staggering.

Indeed, Nusa Tenggara and Maluku together comprise a major centre of global bird diversity. Of the 562 species recorded in the region, 144 of these are endemic. Maluku alone is home to 94 bird species found nowhere else on Earth—more endemic birds than for any other similar area. Mainland Malaysia, on the other hand, has a paltry one per cent endemism.

In Nusa Tenggara, changes in the avifauna continue progressively along the necklace-like chain. Thus Lombok, because it is the first oceanic island in the Nusa Tenggara chain, is strikingly different from Bali on the west, and somewhat different from Sumbawa to the east. Sumbawa and Flores have rather similar avifaunas and the still poorly known avifaunas of Lomblen, Pantar and Alor are evidently derived from Flores, all the while becoming more and more distinct from the mother populations. 'Birds entering from Java,' Wallace explained, 'are most numerous in the island nearest Java; each strait of the sea to be crossed to reach another island offers an obstacle and thus a smaller number get over to the next island ...

... On passing to Flores and Timor the distinctness from the Javanese productions increases and we find that these islands form a natural group, whose birds are related to those of Java and Australia, but are quite distinct from either. ... With the exception of two or three species which appear to have been derived from the Moluccas, all these birds can be traced, either directly or by close allies, to Java on the one side, or to Australia on the other, although no less than 82 of them are found nowhere out of this small group of islands.'

Timor has the highest number of endemic species of any of the other islands of Nusa Tenggara—not surprising because it is also the biggest island of that chain. If the islands surrounding Timor are included, Wetar to the north, Damar and Babar to the north-east, then the number of endemics expands to 23, over 16 per cent: a high degree of endemism. Endemism even extends to generic level with the genera *Buettikoferella* and *Heleia* being endemic.

On Timor, Wallace noticed that endemicity was strongest with birds derived from Australia. 'The larger proportion of birds from Java are identical to Javanese birds,' he said, 'whereas an almost equally large proportion of birds derived from Australia are distinct, though often very closely allied species.' Wallace attributed this to the idea that birds first 'peopled' the islands from Australia. The more likely explanation, however, lies in the continual influx of species from Java and Bali, which flows along the loosely linked islands diluting variation and speciation of Oriental birds.

During the long-lasting Pleistocene glaciations, also, Nusa Tenggara would only have been separated from mainland Asia by very narrow water channels, with lands on either side clearly visible. Indeed the distance from Bali to Lombok via Penida Island at these times may have been as little as a kilometre. The proximity to South-East Asia and the stepping-stone nature of the islands has virtually telescoped species along the chain. The result is that even Timor, which is very much closer to Australia than to Java, has nearly half its birds ultimately derived from Asia.

The constellation of islands of Maluku, on the other hand, has maintained its isolation, even from a Pleistocene-expanded Australia, across water gaps of rarely less than 70 kilometres. Most of its birds are, respectively, either endemic, derived from New Guinea, or derived from the Oriental stock of Sulawesi.

Wallace was particularly transfixed with the glorious birds and insects of Maluku which make it 'the classic ground in the eyes of the naturalist and characterize its fauna as one of the most remarkable and beautiful upon the globe'.

Even the depauperate (relatively low) number of species, typical of oceanic islands, did not dampen his enthusiasm. 'In everything but beetles, these eastern islands are very deficient, compared with the western (Java, Borneo, etc.) and much more so if compared with the forests of South America, where 20 or 30 species of butterflies may be caught every day and on very good days a hundred, a number we can hardly reach here in months of unremitting search. In birds there is the same difference. In most parts of tropical America we may always find some species of woodpecker, tanager, bush-shrike, chatterer, trogon, toucan, cuckoo and tyrant-flycatcher; and a few days active search will produce more variety than can be here met with in as many months. Yet along with this poverty of individuals and of species there are, in almost every class and order, some one or two species of such extreme beauty or singularity as to vie with, or even surpass, any thing that even South America can produce.'

In Banda, Wallace remarked on the handsome fruit-pigeon *Ducula bicolor* whose loud, booming note he heard continually. He discovered a small fruit-dove *Ptilonopus regina* peculiar to Banda. In Ambon, large flocks of the fine crimson lory *Eos borne* formed a magnificent spectacle when they settled down on a flowering tree to feed on the nectar. Here Wallace also obtained a couple of specimens of the fine racquet-tailed kingfisher of Ambon, *Tanysiptera galatea*, 'one of the most singular and beautiful of the beautiful family'.

'These birds differ from all other kingfishers', Wallace explained, 'by having the two middle tail-feathers immensely lengthened and very narrowly webbed, but terminated by a spoon-shaped enlargement, as in the motmots and some of the hummingbirds. They belong to that division of the family termed king-hunters, living chiefly on insects and small land-molluscs, which they dart down upon and pick up from the ground, just as a kingfisher picks a fish out of the water. They are confined to a very limited area, comprising the Moluccas, New Guinea and Northern Australia. About ten species of these birds are now known, all much resembling each other, but yet sufficiently distinguishable in every locality. The Amboynese species ... is one of the largest and handsomest. It is fully 17 inches long to the tips of the tail-feathers; the bill is coral red, the under surface a pure white, the back and wings deep purple, while the shoulders, head and nape and some spots on the upper part of the back and wings, are pure azure blue. The tail is white, with the feathers narrowly blue-edged, but the narrow part of the long feathers is rich blue. This was an entirely new species.'

In Halmahera, Wallace shot a pair of the 'most beautiful little long-tailed bird, ornamented with green, red and blue colours and quite new to me. It was a variety of the *Charmosyna placentis*, one of the smallest and most elegant of the brush-tongued lories'. The island of Morotai, while close to the north-eastern extremity of Halmahera, has its own suite of endemics. A kingfisher *Tanysiptera doris*, a honeyeater *Philemon fuscicapillus* and a large Paradise Crow *Lycocorax pyrrhopterus*, are quite distinct from allied species found in Halmahera. The Paradise Crow is all the more remarkable because it is one of only two cases where the bird of paradise family Paradisaeidae, extends beyond the Sahul Shelf of Australia. Here in Maluku its endemism extends to genus level.

In Bacan, Wallace found parrots and fruit-pigeons which he had already collected on Halmahera but 'was rewarded by finding a splendid deep blue roller (*Eurystomus azureus*), a lovely golden-capped sunbird (*Nectarinea auriceps*) and a fine racquet-tailed kingfisher (*Tanysiptera isis*), all of which were entirely new to ornithologists'.

In Buru, out of 66 species of birds which Wallace collected, no less than 17 were new, or had not been previously found in any island of the Moluccas. 'Among these were two kingfishers (*Tanysiptera acis* and *Ceyx cajeli*); a beautiful sunbird (*Nectarinea proserpina*); a handsome little black and white flycatcher (*Monarcha loricata*), whose swelling throat was beautifully scaled with metallic blue; and several of less interest.'

On Buru, Wallace witnessed something unbelievable. It started when he found himself constantly mistaking two birds for one another. Not an unusual thing normally, except that these two birds, the oriole *Oriolus bouroensis* and the friarbird *Philemon moluccensis*, are from entirely different families and the oriole had abandoned its normally gay yellow tints to adopt the plain brown of the friarbird—it was as if the oriole was mimicking the friarbird. Common enough in insects, this was the first time that a case of visual mimicry had been suggested for birds. On adjacent Seram he found the same thing, but was amazed to discover that it was between other species of oriole and friarbird.

Wallace's claim of visual mimicry in a bird, was indeed not believed.[17] Instead the observation was interpreted as convergent evolution between animals, a common enough occurrence. More than a century later, well-established cases of visual mimicry in birds are scarce. It was not until 1982 that there was any further study on Wallace's observation. At that time, the ecologist Jared Diamond took the case on and confirmed the mimicry. So compelling was it that the two birds had been misdescribed by taxonomists and mislabelled in museums.[17]

Biological mimicry involves three roles: a mimic which resembles another species and thereby gains some advantage for itself; a model, which is imitated by the mimic; and a signal receiver, which mistakes the mimic for the model and thereby becomes the victim of the mimic's deception. So of the friarbird and oriole which is the mimic and which the model? 'We have two kinds of evidence to tell us which bird in this case is the model and which is the copy', Wallace said. 'The honeysuckers are coloured in a manner which is very general in the whole family to which they belong, while the orioles seem to have departed from the gay yellow tints so common among their allies. We should therefore conclude that it is the latter who mimic the former.'

This explanation effectively dismissed convergent evolution since that assumes the evolution of taxa toward one another. In this case there was no evolution of friarbirds toward orioles. The oriole, actually, goes even further, mimicking not only plumage but also posture, movements and flight and even vocal mimicry. What benefit does the oriole obtain from such deception? Wallace's explanation was this:

'The Tropidorhynchi [friar-birds] are very strong and active birds, having powerful grasping claws and long, curved, sharp beaks. They assemble together in groups and small flocks and they have a very loud bawling note which can be heard at a great distance and serves to collect a number together in times of danger. They are very plentiful and very pugnacious, frequently driving away crows and even hawks, which perch on a tree where a few of them are assembled. It is very probable, therefore, that the smaller birds of prey have learned to respect these birds and leave them alone and it may thus be a great advantage for the weaker and less courageous *Mimetas* [oriole] to be mistaken for them.'

Since bird-eating hawks are exceptionally rare on Buru and Seram, Wallace's observation lost much of its credibility. The oriole's mimicry, in fact, is much more deceptive.

Orioles, friarbirds and other honeyeaters have similar dietary preferences and are commonly seen together in the same trees, a fact which inevitably aggravates the friarbirds. The largest of the honeyeaters and known generally for their aggression no matter what size they are, friarbirds are notorious bullies. But even the prickly friarbird has a soft spot for its juveniles and it is the plumage of the smaller juveniles, in particular, that the oriole mimics. In what can be seen as biological poetic justice, the fiery friarbird is at once the model and the victim of the deceit. The oriole, by resembling a larger bird, may also succeed in scaring off smaller species by appearance alone. So precise is the mimicry that where orioles are in fact larger than friarbirds, such as in northern Australia, no mimicry occurs since the bigger oriole is able to turn the tables and bully the friarbird.

On Maluku, virtually every island became a treasure island for Wallace. He was constantly overwhelmed by new discoveries of birds 'of gay and delicate colours, adorned with the most gorgeous plumage' and dazzled by the large and showy butterflies. That was until he went to Seram.

'All the time I had been in Seram I had suffered much from the irritating bites of an invisible acarus, which is worse than mosquitoes, ants and every other pest, because it is impossible to guard against them. This last journey in the forest left me covered from head to foot with inflamed lumps, which, after my return to Amboyna, produced a serious disease, confining me to the house for nearly two months—a not very pleasant memento of my first visit to Seram.'

Wallace had walked six days to the centre of Seram in steep, densely forested country. So rugged was the walk that his two pairs of shoes fell to pieces and, in his 'stockings', he finally limped back, virtually lame. Apart from some new butterflies, the reward for his efforts was a debilitating disease. 'I saw not a single ground-thrush, or kingfisher, or pigeon', he said, 'and, in fact, have never been in a forest so utterly desert of animal life as this forest appeared to be. ... it was a complete puzzle which to this day I have not been able to understand.'

The answer to the puzzle may lie with Wallace himself and his focus on the peculiar and new. Seram is known to have at least 225 butterfly[67] species and 124 species of breeding birds,[8] not an inconsiderable number, even for a depauperate island fauna. However, in a region known for its endemism, which reaches up to 100 per cent for birds on some islands, only 10 per cent of Seram's birds are endemic[8] and only two per cent of its butterflies are endemic[67]—compared with 7 per cent for the nearby smaller Buru. An explanation for the puzzle probably lies in the age of the two islands. While both have a similar geological age, it is likely that Buru has been above water for longer than Seram, allowing more time for the evolution of peculiar species. The high level endemism of Morotia and Bacan (which has the other endemic bird of paradise, the Standard-wing Bird of Paradise *Semioptera wallacii*) indicates that these islands are older than the others in the group.

Wallace always seemed to know which birds were to be considered Australian and which were considered Oriental. Up until 10 years ago, most modern-day ornithologists could not claim to have the same confidence. Wallace's observations, correct to this day, were based on objective analysis of the distribution of bird species and not clouded with the absurd idea that Australasia was some kind of evolutionary antipodean outpost; an outpost which could not possibly have its own unique bird fauna. Despite almost crushing observations against it, the view that Australasia received its birds via five successive waves from superior Eurasian stock, over a 30 to 40 million year period, was held by some until 1986.

It was thought that in the first wave, pulsing through Indonesia about 40 million years ago, came the ancestors of the emu, cassowary, megapodes, wood-swallows, parrots and honeyeaters. The second wave included ancestors of the plains-wanderer, fruit doves, birds of paradise and figbird. Surfing in on the third wave came the Jacana, kookaburras, paradise-kingfishers, flycatchers, fantails, monarchs, robins, treecreepers and the Metallic Starling. In the fourth wave came the Australasian Grebe, Rufous Night Heron, Spotted Harrier, Australian Hobby, Australian Kestrel, button-quails, owls (sooty and masked), Sacred and Collared Kingfisher, Red-backed and Forest Kingfishers, White-bellied Cuckoo-shrike, Olive-backed Oriole and crows. In the last wave came species not differentiated from Asian species.

Even drifting continents could not shift this view. The fact that Australia lay thousands of kilometres to the south of Indonesia for most of this time was incidental. Austral warblers, flycatchers, thrushes, babblers and other Australian passerines (songbirds) were merely Asian look-alikes. Since these songbirds were considered to have arrived in Australia about 30 million years or so ago, it was thought that songbirds worldwide had evolved and radiated around then, spilling over, finally, into an empty Australia.

There were a few problems with this view of the world. For a start, since over 90 per cent of the Australasian land and freshwater bird species are unique to the region, what had happened to the Asian representatives after each wave? Like a heart, the

Fig 1. Wallace's selection of animals from Papua New Guinea contrast with those from a Bornean forest (Fig 8). Most obvious is the lack of mammals, the only example a marsupial. Left to right and top to bottom: a tree kangaroo *Dendrolagus inustis*, the Fairy Lorrikeet *Charmosynpa papou*, the Twelve-wired Bird of Paradise *Seleucides melanoleuca*, the Common Paradise Kingfisher *Tanysiptera galatea* and the Western Crowned Pigeon *Goura cristata* (*coronata*).

Fig 2. At age 39, Alfred Russel Wallace had explored the depths of the Amazonian jungles, and the length and breadth of the remote and exotic 'Far East'. This young 'fly-catcher' had collected 125 660 specimens of natural history, including many species unknown to science. A self-educated drifter without prospects of regular employment, had invented the field of biogeography and used it to develop and then test his famous Theory of Evolution, which, in a breathtaking coincidence, was developed simultaneously with, but independently of, Charles Darwin's.

Fig 3. The characteristic birds of a Malaysian forest have their evolutionary roots in Eurasia and Africa, and are as different to Australian birds as pangolins are to possums. Today Australian and Eurasian birds meet in the vicinity of The Wallace Line.

'From a giant pitcher plant a man could escape only with the help of a friend'. Illustration in S. Leonard Bastin's *Scientific American* article, 1909.

'A goat-eating butter-wort of the future'. Illustration in S. Leonard Bastin's *Scientific American* article, 1909.

Figs 4 and 5. During Wallace's time carnivorous plants were unheard of. Had he known, he might have thought twice about slaking his thirst with the digestive juices of the exotic, and carnivorous, pitcher plant. The concept of carnivorous plants was greeted with disbelief when introduced to the world by Darwin in 1865. In Victorian England, plants were meant to be passive things. Even in 1909, years later, the reaction to carnivorous plants was still hysterical.

Illustration by Martin Woodcock in Alan Kemp's *The Hornbills*, 1995, by permission of Oxford University Press.

Fig 6. The nesting female hornbill locks herself away, sealing the entrance of an appropriate tree hole so that a narrow vertical slit becomes her lifeline to the outside world. Sanitation in such a cosy home is dealt with in a most hygienic fashion. The female turns away from the entrance, reversing to the narrow opening, carefully positions her anus and then defecates with considerable force sending out a spray of material.

'View of Krakatoa During the Earlier Stage of the Eruption From a Photograph Taken on Sunday the 27th of May.' Lithograph from the Report of the Krakatoa Committee of the Royal Society, *The Eruption of Krakatoa and Subsequent Phenomena*, 1888. (courtesy Mitchell Library, Sydney.)

Fig 7. Krakatau went into paroxysm in August 1883. Tsunamis radiated out from the volcano at speeds of up to 800 kilometres per hour and reached 40 metres high. They killed 36 360 people. Two hundred and twenty kilometres from the area, carcasses of wildlife, including tigers, were found floating among the pumice. The dust that was lifted into the stratosphere circled the globe and affected world temperatures for three years. Krakatau was by no means the largest volcanic eruption in a land of global-scale fireworks, but it is perhaps the world's most famous. The first year of mass media was 1883, when the 'modern' global telegraphic communication transmitted the eruption to the world.

Fig 8. Wallace's selection of animals from Borneo contrast with those from a New Guinean forest (Fig. 1): all mammals are placental and there are no birds. Left to right and top to bottom: The Western Tarsius *Tarsius bancanus*, one Flying Lemur *Cynocephalus variegatus* in flight and another seated, the Pentail Treeshrew *Ptilocerus lowii*, the Malay Tapir *Tapirus indicus* and two Lesser Mousedeer *Tragulus javanicus*.

Wood engraving in Wallace's *The Geographical Distribution of Animals*, 1876 (courtesy Australian Museum, Sydney).

Frontispiece in Wallace's *The Malay Archipelago*. (courtesy Mitchell Library, NSW).

Fig 9. All the great apes of the world except Humans are now endangered. For most of this century the Orang-utan was considered the least interesting of the apes, since it appeared to lack the intelligence of the chimpanzee and the Tarzan-born brutality of the gorilla. The Orang-utan depicted here was acting in self-defence. A few minutes before this moment it had been feeding innocently on the young shoots of a palm by a Bornean river side.

Wood engraving in Wallace's *The Malay Archipelago*, 1869 (courtesy Mitchell Library, NSW).

Fig 10. Even today a voyage to Aru is considered a wild and romantic voyage. Wallace was the first European to visit this remote island in search of the legendary birds of paradise. The Greater Bird of Paradise, seasonally participates in extraordinary 'dancing-parties' within a specific communal display area, known as a 'lek'. This habit enabled the natives of Aru to ambush them with blunt arrows which would stun the bird so that it would drop to the ground. On the ground the bird was secured and killed without a drop of blood tarnishing the feathers.

pulsing flow of birds into Australia must have been through some kind of one-way valve. This valve must have regularly suffered some sort of embolism to stop the flow, and then a massive infarction to cause the wholesale extinction of any surrounding species. Clearly it was time for some common sense.

This came in 1986, when Charles Sibley and Jon Ahlquist presented a revolutionary phylogenetic, or family, tree of the world's birds. Drawn on a poster several metres long it became known as the 'tapestry'.[50] Representing more than an decade's work of molecular systematics on 1700 species of birds it is the only molecular family tree of any large group of animals.

Sibley and Ahlquist compiled the birds' family tree by using a DNA–DNA hybridisation technique to measure genetic distances among bird species. This complex technique compares strands of DNA between two individuals. Firstly DNA is extracted from two different birds of different species. The DNA is 'shorn' into small fragments. DNA from one bird is then heated until its double-helix structure melts and unzips into single strands. These single strands are then radioactively labelled and become known as the 'tracer'.

The DNA from the other bird is also melted into single strands. It is then mixed in great excess—usually a thousand fold—with the tracer DNA. This mixture is then incubated. Under low temperatures the DNA strands will tend to zip up. Because there is so little 'tracer' DNA most of it will reanneal with the other bird's DNA, rather than with its own complementary strands. The principle of the technique is that differences between the DNA of the two species will determine how precisely they zip.

The hybrid DNA is then placed into a solution which binds only the newly zipped-up, double-stranded combination DNA. This solution is heated slowly. At the lowest temperatures, poorly complemented fragments—those of two species distantly related—will melt and unzip. The amount of single-stranded

DNA washed off at each temperature, measured by its radioactivity, is plotted on a graph. By comparing graphs of hybrids with graphs where the DNA comes from the same species— where the temperatures required to melt the DNA are high—a genetic distance can be calculated, which provides an estimate of how closely related the two species are.

Using this information plus the knowledge of plate tectonics, a more convincing story of the birds has emerged. The early birds, newly differentiated from their ancestral reptile stock, first began to occupy various ecological niches throughout the world in the late Jurassic and earliest Cretaceous, around 130 million years ago. From fossil feathers, birds are known to have been present in Australia at least as early as 120 million years ago, even before the first flowering plants migrated around the margins of Gondwana and at the same time as major rifting commenced between Africa and South America.

Australia's songbirds evolved early, somewhere in eastern Gondwana. Molecular biology produced some surprises. For instance, it showed that the world's crow family consists not only of crows, ravens, rooks, jackdaws and jays, but also nutcrackers, birds of paradise, Australian magpies and currawongs, butcherbirds and woodswallows, the Borneo Bristlehead and the Old World orioles and the cuckoo-shrikes and trillers. They represent an ancient, endemic, basically Australasian group which not only radiated far beyond Australia but, more recently (geologically speaking), recontributed members back into Australia, such as the genus *Corvus*, which contains Australia's better known crows, and the orioles.

Northern hemisphere fossil evidence, by the absence of songbird fossil in otherwise rich fossil beds of birds, at least tenuously supports these findings. Also absent are groups such as pigeons and parrots, both found in abundance in Australia. It is possible that all of these may have originated in the southern hemisphere and migrated northward after 55 million years ago. It is indeed looking increasingly possible that all the world's songbirds originated in Gondwana and later migrated into Europe. Aussie songbirds, at any rate, are as Australian as kangaroos.

Endemism in the birds of Nusa Tenggara and Maluku is dominated by a relatively small number of families: pigeons with 15 endemic species; parrots with 16 endemic species; honeyeaters with 15 endemic species; and white-eyes with 13 endemic species. In addition, Wallace found a preponderance of kingfishers and sunbirds. These are all families which have a remarkable capacity for colonising oceanic islands and then speciating.

In complete contrast, mammals show a remarkable incapacity for colonising islands, a fact frequently enough remarked upon by Wallace. No Australian marsupial has naturally survived the wide water crossing to Timor. Even the marsupial cuscus *Phalanger orientalis* that Wallace observed there, was likely to have been brought to Timor by people, since it is widespread and has the appealing habit of producing twins.[19] Lombok's impressive list of mammals, too, includes the Long-tailed Macaque, Silvered Leaf Monkey, Pangolin, Plantain Squirrel, Sunda Island Porcupine, Leopard Cat, Little Civet, Common Palm Civet, Barking Deer and the Sunda Shrew. All of these are known to be human commensals, kept as pets or to thrive in human-disturbed environments[37] and are, therefore, likely to have been transported to Lombok by people. Without these animals, Lombok's mammal population would dwindle down to a population of bats and rats. The small permanent gaps of five to 20 kilometres, such as those which separate Borneo's northern offshore islands from the oceanic islands of the Philippines, will separate a rich mammalian fauna from an oceanic one.

Unfortunately, there is a singular lack of knowledge of the

mammals of island faunas of Nusa Tenggara and Maluku. Cursory visits or, at most, just a few weeks of field work[19] provide the sum total of knowledge. And even this has occurred mostly in the last decade. Even where mammals are relatively well known, such as on Lombok, it is difficult deciding which animal is native and which is not. The species composition of mammals on most islands has been grossly altered by people.

Perhaps reflecting this, Wallace rarely discusses the distribution of mammals apart from curiosities and human commensals. Fruit-bats, in particular, show a high degree of endemism among oceanic islands. As a group, they are most diverse in the South-East Asian and Australo–Pacific region and, like songbirds, may well have evolved in this region.[36] Wallace's account of them, rather disparagingly, is restricted to their culinary attributes.

'They [the Portuguese Protestants of Bacan] are almost the only people in the Archipelago who eat the great fruit-eating bats called by us 'flying foxes'. These ugly creatures are considered a great delicacy and are much sought after. At about the beginning of the year they come in large flocks to eat fruit and congregate during the day on some small islands in the bay, hanging by thousands on the trees, especially on dead ones. They can then be easily caught or knocked down with sticks and are brought home by basketfuls. They require to be carefully prepared, as the skin and fur has a rank and powerful odour; but they are generally cooked with abundance of spices and condiments and are really very good eating, something like a hare.'

Mammals which have successfully dispersed to oceanic islands also show a remarkable capacity to speciate. In the oceanic islands of the Australasian region, for instance, there are 80 species of fruit-bat of which an extraordinary 73 per cent are endemic.[19] In Nusa Tenggara, discoveries, largely by Western Australian Museum scientists, led by Darryl Kitchener, over the past decade amount to no less than 12 new species of bats and rats.[37] And of the known mammalian fauna of these islands, at least 10 bats and seven murid rodents (rat-like rodents) are known to be endemic at the species level. Sulawesi and Flores, in particular, have large numbers of endemic genera of rodents. Tellingly, the speciation of rodents, at least to generic level, has been particularly rapid from about 8 million years ago, coinciding with the increasing number of island arcs appearing in the region.

The patterns of speciation in Nusa Tenggara at the species level reflect the Pleistocene configuration of islands. During the last major glacial maximum four main islands existed: Lombok to Sumbawa; Komodo to Adonara; Timor, Semau and Roti; and the Tanimbar islands. Overlaying this is a more modern pattern of sub-speciation with different island species differing in appearance from each other.

In a region of surprises, the recent work of scientists has turned up another surprise. It seems that the much maligned rats of the genus *Rattus*—such as the Black Rat *Rattus rattus*, the one that introduced plague to Europe in the Middle Ages via its fleas—actually evolved in South-East Asia,[75] possibly on its eastern oceanic islands. The spread of *Rattus* to Europe probably occurred not long after the return of the first explorers to Europe. Indigenous species within *Rattus* are found in Australia, New Guinea, China, India, the Philippines and Europe. However, the bulk of the species in *Rattus* are South-East Asian and all genera most closely related to it are also South-East Asian. It seems that the genus *Rattus* evolved relatively recently with secondary centres of evolution in Sulawesi, New Guinea and Australia. It has only recently spread to mainland South-East Asia and Australia. Indeed Australia's own native rats are closely related to *Rattus norvegicus*, the Brown or Ship Rat.

More isolated Maluku reflects the more accidental or sweepstakes nature of colonisation. Halmahera, for instance, has only one native rodent, while Seram has five endemic species, as well as an endemic bandicoot *Rhynchomeles prattorus*, the Northern Common Cuscus *Phalanger orientalis* and the Common Spotted Cuscus *Spilocuscus maculatus*, though both these latter marupials were probably introduced to Seram by people.

Like birds, the relationships of Moluccan mammals lie mainly with New Guinea. While numbers of species are low, endemism among the mammals of Maluku is high. In Halmahera, for instance, a 3500-year-old fossil record of three or four marsupials (bandicoot, sugar glider and cuscus) and a native rat, shows that the non-flying mammals of the island were all Moluccan endemics or New Guinean species.[19] Most of the 20 or so fossil bats also show this relationship although a significant element originated from Sulawesi; a pattern also reflected in birds.

In Bacan, Wallace discovered the marsupial *Cuscus ornatus*, the distribution of which spreads to Morotai and Halmahera. Apart from two endemic species of cuscus found on islands Wallace did not visit (the Obi Cuscus *Phalanger rothschildi* and the Gebe Cuscus *Phalanger sp.*) his description of the marsupial fauna of Maluku is quite accurate. 'The first [of the marsupials] is the small flying opossum (*Belideus ariel*) [*Petaurus breviceps*], a beautiful little animal, exactly like a small flying squirrel in appearance, but belonging to the marsupial order. The other three are species of the curious genus *Cuscus*, which is peculiar to the Australo–Malayan region. These are opossum-like animals, with a long prehensile tail, of which the terminal half is generally bare. They have small heads, large eyes and a dense covering of woolly fur, which is often pure white, with irregular black spots or blotches, or sometimes ashy brown with or without white spots. They live in trees, feeding upon the leaves, of which they devour large quantities. They move about slowly and are difficult to kill, owing to the thickness of their fur and their tenacity of life. A heavy charge of shot will often lodge in the skin and do them no harm and even breaking the spine or piercing the brain will not kill them for some hours. The natives everywhere eat their flesh and as their motions are so slow, easily catch them by climbing; so that it is wonderful they have not been exterminated.'

Curiously, most of the endemic marsupials, particularly the most distinctive ones, are found in montane forests. The same trend is found in birds, where most of the old endemic birds are montane. Even in Borneo, Sumatra and Java, a significant proportion of endemic birds is montane. This is because mountains are themselves evergreen islands even during the drier glacial periods. Any speciation that occurs in the lowlands in the intervening wet periods, like those of today, are swamped during the glacial periods, whereas the mountains remain islands in the sky. These are discussed in the next chapter.

'The almost entire absence of Mammalia and of such widespread groups of birds as woodpeckers, thrushes, jays, tits and pheasants' convinced Wallace that he was 'in a part of the world which has in reality but little in common with the great Asiatic continent, although an unbroken chain of islands seems to link them to it.'

If Wallace had followed his own logic he would have concluded that the region equally had but little in common with the great Australian continent. For instance, eastern Indonesia does not have such widespread groups of birds as fairy wrens, sitellas and treecreepers, whipbirds, wedgbills and quail-thrushes. In common with the woodpeckers, thrushes, jays, tits and pheasant, they have a continental heritage.

The strings and constellation of the islands in the sea between Australia and Asia are neither Australian nor Asian but literally a biogeographic law unto themselves. The irony is that Wallace—who perhaps more than anyone else in history, emphasised their singular nature, eulogised their dazzling beauty and accurately explained their fauna's island origins—still concluded that the region was merely an intergradation zone between the Australian and Oriental biogeographic region; like some second-rate continental appendage.

CHAPTER 7

ISLANDS IN THE SKY

I N AUGUST 1854, Wallace stepped out of the cool, dark forest onto a steep slope of bare rock extending along the mountain-side as far as he could see. From cracks and fissures in the rock erupted a luxuriant vegetation of ferns and *Dacrydium* conifers. Among these, pitcher plants twined, their curious pitchers of various sizes and forms hanging abundantly. Sweating and thirsty from the exertion of climbing several kilometres up the steep sides of Mount Benom, in the middle of the

Malay Peninsula, Wallace looked about in vain for the promised spring. Finally, he turned to the pitcher plants to slake his thirst. Each of the natural jugs was indeed full of about a litre of liquid, but it was a virtual soup of dismembered and dissolved pieces of insects. Wallace drank, spitting out the bigger bits, though he found it was surprisingly palatable.

Wallace was to come across the *Nepenthes* pitcher plants again, mostly on mountain tops, where, he said, they reached their greatest development 'running along the ground, or climbing over shrubs and stunted trees, their elegant pitchers hanging in every direction. Some of these are long and slender ... others are broad and short. Their colours are green, variously tinted and mottled red or purple. The finest yet known were obtained on the summit of Kini-balou, in north-western Borneo. One of the broad sort, *Nepenthes rajah* will hold two quarts [2.27 litres] of water in its pitcher. Another, *Nepenthes edwardsiania* has a narrow pitcher 20 inches long, while the plant itself grows to the length of 20 feet'.

The tropical pitcher plant *Nepenthes* has at least 70 species in South-East Asia and northern Australia. Intriguingly, they are also found in Madagascar and Sri Lanka, a distribution suggestive of a Gondwanan origin. Wallace thought them a 'marvellously curious vegetable production'.

At the time, he did not know that they were carnivorous. The fact of plant carnivory was revealed to the world by Charles Darwin in 1865, ten years after Wallace had drunk his soup of insect-steeped fluid. Darwin's 1875 work, *Insectivorous Plants*, did not go down all that well in Victorian England. Plants were meant to be passive and peaceful things. A role as fierce predators in the animal kingdom was not at all a fitting one for plants. It was quite unnatural. In 1881, a criticism even appeared in the learned journal *Scientific American* which said that plants were not carnivorous in any sense of the word and that Darwin's theories should be taken *cum grano salis* (with a grain of salt).[12]

Nevertheless, the Victorians quickly recovered from the impropriety of the concept and, rather breathlessly, began instead to acquire them. No glasshouse was complete without its *Nepenthes* hanging from the rafters. Like voyeurs, wealthy Europeans fed them insects, watching them slip on the sleek sides of the pitcher and plunge into the digestive pit.

Carnivorous plants had captured the gruesome imagination. In 1909, in another article in *Scientific American*, writer S. Leonard Bastin[12] speculated on whether house plants would some day attack human beings and take over the world. He was concerned that more and more plants were becoming carnivorous and that something should be done about it. He speculated that the carnivorous 'habit' could not only become more widespread, but that the present carnivorous plants would become larger. The article, rather hysterically, depicted giant sundews snaring large birds such as storks; butterworts that could hold down and dissolve goats; large underwater bladderworts that devoured crocodiles; and man-eating pitcher plants!

In truth, pitchers are pitfall traps. Like children to the gingerbread house, insects are attracted to the trap by means of nectar, advertised by the vivid colouration of the pitcher. The most abundant nectar is found just within the entrance. To sip the sweetest nectar the insects wander over a slippery surface. They suddenly lose their footing and plummet to the bottom of the abyss, from which escape is impossible.

Extraordinarily, the *Nepenthes* pitchers are only leaves, albeit highly modified. The *Nepenthes* plants are characteristically tropical vines which can reach high into the jungle canopy from the forest floor. The pitchers form at the end of the leaves, first as a tendril growing from the leaf tip. A noticeable swelling at the end of the tendril is the first sign of an incipient pitcher. As the swelling expands, the tendril hangs down due to the increased weight. The tendril bends, like a joint, at the base of the pitcher, forcing it upright.

As the flat pitcher approaches maturity it is suddenly inflated with air. Microscopic glands pitted on the inside walls of the pitcher then secrete digestive fluid until the pitcher is charged with liquid. Colour infuses the closed pitcher. After a few days the lid opens. The pitcher is operational. The lid has a dual role. It forms a rain-protecting canopy over the mouth of the pitcher, stopping the pitcher from filling with rainwater, thus overflowing and losing its nutritious liquid. It is also a device of seduction. Deeply flushed with colour, it attracts insects which come to sip the nectar oozing from glands which liberally cover its inner surface.

The pitcher mouth has a corrugated rim which is hard, glossy and rounded. The parallel ribs which form the corrugations each terminate within the pitcher in a sharp downward-pointing tooth. Deceptive, single, nectar-secreting glands are found within the curves between the teeth. Beneath the rim and its teeth, but above the glassy surface of digestive juice, is a smooth and waxy zone. This usually extends about one-third of the way down the pitcher. Insects alighting on the pitcher, or on its nectar-rich lid, tend to pace about in search of a more abundant source of nectar and sooner or later will find their way to the drops of nectar between the down-pointing teeth of the inner rim. The rim generally provides too limited a space to accommodate all their feet and they seek a foothold on the more spacious waxy zone. Like walking on thin ice, the wax easily breaks away, clogging insect footpads and sending the animals plunging into the pit.

Insects, small reptiles, amphibians and even mammals die by drowning in the pitcher. Their struggling activity stimulates the digestive glands of the pitcher, which then secrete acids and enzymes heavily, the fluid becoming acid in a few hours. Digestion takes hours or days depending on the size of the victim. The fluid of an unopened pitcher is generally neutral, but shortly after suitable food is added, acidity increases. This decreases during digestion so that the fluid once again becomes neutral. Digestive enzymes known to be present are ribonuclease, lipase, esterase, acid phosphatase and protease.[60] Ever the optimist, Wallace's palatable beverage was little more than pitcher plant-pepsin.

The shape of a plant's pitchers changes depending on the location of the pitcher. The leaves closest to the ground have short tendrils while the leaves higher up have increasingly longer tendrils. These lasso around branches as the vine climbs up through the canopy. In the process, the pitcher also changes. The lower pitchers, squatting on the ground, are broad and tub-shaped; their prey, crawling insects such as ants and cockroaches. They have a broad rim and often a large spur at the hinge of the lid and rim and two 'wings' running down the front of the pitcher. The wings are fringed and face the front of the plant.

Upper pitchers dangle in the air on the end of their tendrils. In these pitchers, the wings are thinner and less fringed and the pitcher shape is elongated. The pitcher itself faces away from the plant. As the pitcher plant grows even more, the distance between each leaf increases, the leaf becomes thinner and the pitcher becomes tapered toward the base. Deadly decorations hanging in mid-air, they attract flying insects such as moths and wasps. At this stage, when the plant is from one to four metres tall, the flower will form. The flowers, as drab as the pitcher is stunning, are small, about three millimetres across, of a green-bronze or reddish shade, with many borne on a central spike.

Nepenthes grow in infertile ground, often in peats and bogs which characterise the tropical-alpine environments of mountain tops. Here *Nepenthes* grow among the evocatively named 'elfin forests'. These mysterious forests of stunted and twisted trees, draped in garlands of moss and beards of lichen, grow in infertile soils with slow rates of litter decay and subsequent humus mineralisation. Nutrient analysis of their leaves, which to some extent reflects that in soils, shows that phosphorus is particularly deficient. In complete contrast, analysis of the pitcher plants shows nearly three times as much phosphorus and twice as much potassium. Clearly, supplements of flesh provide them with a more balanced diet. Interestingly, *Nepenthes* that are overfertilised produce fewer pitchers or no pitchers at all.

In Java, Wallace did something uncharacteristic. He took a walk up a mountain to look, not at animals, but at vegetation. He had read that tropical mountain environments changed with height—from tropical lowlands to temperate highlands. It was like walking from the equator to the poles. The change started with the lower montane zone emerging from a sea of lowland vegetation. The trees in this zone are shorter and rarely have buttresses or cauliflory (the ability to bear flowers and fruit on the trunk). Climbers are also rare, but certain epiphytes like orchids are common. Higher, in the upper montane zone, the trees are squat, crooked and gnarled. The leaves are thick and small and no longer have drip tips to shed water. Creepers are rare and so are orchids. These 'elfin forests' are enshrouded in mist and draped in garlands of moss and beards of lichen. Higher still, the subalpine forest is a complex of grassy, heathy and boggy areas, thick with carpets of liverworts, moss and lichen. If anything, these peaks resemble Scottish moors. Some of the plants are even in the same genus.

In Sulawesi, for instance, the characteristic plants of the high mountains are *Rhododendron*, *Vaccinium* (cranberry or bilberry) and wintergreen *Gaultheria*. The *Rhododendron* forms small trees with tightly stacked, thick leaves. Beneath these are found small, shrubby wintergreen and colourful flowering herbs such as daisies, ginger *Alpina*, violets *Viola*, *Styphelia* with small pink berries and blackberries *Rubus*. Ragwort *Senecio* and cushion plants are also common. In the open areas, many beautiful herbs can be found in genera typical of temperate regions. Other plants growing in the upper montane zone are magnolias, maples *Acer*, oak *Lithocarpus*, chestnut *Castanopsis* and the endemic *Macadamia hildebrandii*. At the summit of Mount Roroka Timbu are also *Leptospermum* and the ancient Gondwanan conifers from the genera *Dacrycarpus*, *Phyllocladus* and *Podocarpus*, all covered with thick cushions of moss. Here pitcher plants are in their element in the infertile and water-logged soils, winding through the giant mosses and the *Sphagnum* peat mosses and climbing in the 'elfin forest'.

Violet, cranberry, rhododendron, foxglove, honeysuckle, wormwood, wood-sorrel, oak. Nothing could be more different from the three-dimensional architecture of a lowland tropical rainforest than these plants of the highlands evoking a Scottish heather. Such an extraordinary contrast was worth looking at and so Wallace engaged two 'coolies' to carry his baggage and started his climb to the 3000-metre volcanic peaks of Mounts Pangrango and Gede, south of Jakarta.

The first kilometre or so was over open country, which brought him to the forest of 'grand virgin vegetation' which cloaked the mountain to a height of about 1500 metres. Here, he said, 'the road became narrow, rugged and steep, winding zigzag up the cone, which is covered with irregular masses of rock and overgrown with a dense, luxuriant, but less lofty vegetation'. Boiling water gushed down the mountain, foaming and sending up steam as it rushed over its rocky bed.

'At 5000 feet [1500 metres] I first saw horsetails (*Equisetum*), very like our own species. At 6000 feet [1800 metres] raspberries abound. ... At 7000 feet [2100 metres] cypresses appear and the forest-trees become reduced in size and more covered with mosses and lichens. From this point upward these rapidly increase, so that the blocks of rock and scoria that form the mountain slope are completely hidden in a mossy vegetation. At about 8000 feet [2500 metres] European forms of plants become abundant. Several species of honeysuckle, St

John's-wort and guelder-rose abound; and at about 9000 feet [2700 metres] we first meet with the rare and beautiful cowslip *Primula imperialis*, which is said to be found nowhere else in the world but on this solitary mountain summit. It has a tall, stout stem, sometimes more than three feet high, the root-leaves are eighteen inches long and it bears several whorls of cowslip-like flowers, instead of a terminal cluster only. The forest trees, gnarled and dwarfed to the dimensions of bushes, reach up to the very rim of the old crater, but do not extend over the hollow on its summit. Here we find a good deal of open ground, with thickets of shrubby artemisias and gnaphaliums, like our southernwood and cudweed, but six or eight feet [about 2 metres] high; while buttercups, violets, whortleberries, sow-thistles, chickweed, white and yellow cruciferae, plantain and annual grasses everywhere abound. Where there are bushes and shrubs, the St John's-wort and honeysuckle grow abundantly, while the imperial cowslip only exhibits its elegant blossoms under the damp shade of the thickets.'

Near volcanic craters the soil is acid, sterile and baked. The air is toxic with the smell of rotten eggs (hydrogen sulphide); and water flowing out from volcanic craters is little more than streams of sulphuric acid. Remarkably, some plants favour these environments. On the gaping crater of the 3800 metre high volcanic Mount Kerinci in Sumatra, for instance, are grasses, sedges, a daisy *Senecio sumatrana*, a small fern *Gleichenia arachnoidea* which forms prickly thickets, the purple-flowering *Melastoma* and pandans. The magnificent daisy *Anaphalis javanica*, commonly called 'edelweiss', is one of the few plants confined to volcanos. A long-lived pioneer of volcanic ash screes and crater soils, this triffid-like daisy has adapted to the harsh conditions, not by shrinking like most other plants, but by growing to a gigantic eight metres, with a stem as thick as a weight-lifter's wrist. This was the 'six or eight foot' southernwood and cudweed that Wallace mentions.

Wallace was amazed. 'The fact of a vegetation so closely allied to that of Europe occurring on isolated mountain peaks, in an island south of the equator while the lowlands for thousands of miles around are occupied by flora of a totally different character, is very extraordinary.'

He was able to identify many of the plants because they were indeed relatives of those found in his temperate European homeland. Interestingly, the plants that Wallace found on top of tropical mountains were from the most widespread genera of Europe. These ecological 'waifs' are species well-known as pioneer plants of open habitats, having good dispersal ability and wide ecological tolerance and displaying rapid growth and an ability to produce abundant seeds; essentially the same characteristics as weeds.

The migration pathways of some of these waifs are clear. For example *Haloragis micrantha* is found from New Guinea to Japan to the top of the Bay of Bengal. Conversely, the stately herb *Primula prolifera*, with up to five layers of bright yellow flowers, occurs in the high mountains of Sumatra and Java and in the Himalayas, but not anywhere in between. Revealingly, it was itinerants such as these that were the earliest plants to colonise the moraines of north-western Europe during the last glacial—a fact not lost on Darwin.

He said that during the last glacial, these hardy, temperate forms of plants would have extended into the tropics at lower altitudes and during the interglacial would have retreated up the mountains. In the New Guinean highlands, for instance, tropical-alpine vegetation currently occupies a mere 800 square kilometres. At the time of the last Ice Age maximum (18 000 to 15 000 years ago), however, temperatures in the mountains may have been seven degrees Celsius lower than today. Under these conditions, glaciers, currently at a height of around 4500 metres, extended down to about 3200 metres, squashing the upper forest limit down to an altitude of 2000 to 2250 metres and allowing the tropical-alpine flora to occupy an area of about 55 000 square kilometres—seventy times its former range.[62,63]

With the expanded montane vegetation zones lower and closer, dispersal of temperate taxa would have amounted to a hop and step, compared with the jump of today's climatic environment.

The great change of conditions, since the last glacial, has allowed the plants to become modified into new species. The endemicity is extremely localised, often to individual mountain peaks. Wallace, on Mount Pangrango, wrote of 'the beautiful cowslip *Primula imperialis*, which is said to be found nowhere else in the world but on this solitary mountain summit'. On top of 4510-metre Mount Wilhelm in New Guinea, 73 per cent of the species of the widespread genera are endemic. In Sulawesi, the figure is an extraordinary 81 per cent: of the 24 species of *Rhododendron*, 19 are endemic; 13 of the 16 species of *Vaccinium* (cranberry or bilberry) are endemic; and both species of the wintergreen *Gaultheria* are endemic.

The endemism rarely extends to generic level, however, reflecting the recent colonisation of these islands in the sky. For instance, of the 107 native genera on top of Mount Wilhelm, only 10 are endemic to the region; 11 are northern temperate; 28 are southern temperate; and 58 are widespread in both temperate zones.[62]

The flora of the tropical mountain peaks, then, seems to be little more than a collection of opportunists. Even ancient Gondwanan genera, *Dacrycarpus*, *Drimys* and *Pittosporum*, which occur on Mount Wilhelm are opportunistic plants of the montane forests which have waited for millions of years for the opportunity to colonise new mountain environments.

In fact, the only biogeographic boundary for the flora of tropical mountain peaks is an absence of mountains. New montane environments, like new islands, are colonised by such opportunists which, separated from their mother population, then undergo a phase of rapid evolution. In many respects, then, the creation of new mountains in the collision zone between Australia and Asia, combined with the Pleistocene glaciations, have acted as an elixir of youth for many taxa.

The diversity of environments of tropical mountains generally—from the ancient, highly endemic lowland forests through to the recent vegetation of the tropical-alpine peaks—together with climatic fluctuations, has resulted in them being botanical treasurehouses; richer than in any other areas of comparable size in the world.[78]

The jewel of them all is Mount Kinabalu in north-western Borneo. It, indeed, has the richest and most remarkable assemblage of plants on the planet.[52] The highest mountain between the Himalayas and New Guinea, the first botanical stash from Mount Kinabalu, in 1894, amounted to 364 species; 140 of them new species. Today we recognise 4000 species—and only a fraction of the whole mountain has been explored.

Mount Kinabalu is a geologically recent massif, a gargantuan chunk of granite thrust into the surface of the earth, not more than nine million years ago. It was uplifted only in the Pleistocene, within the last two million years. At 4101 metres, Mount Kinabalu is still rising 0.3 centimetres per year,[52] creating a great altitudinal and climatic range, from tropical rainforests near sea level to alpine—even glacial—conditions at the summit.

In the short time during which it has been elevated, numerous new habitats, like freshly sown seedbeds, have been created, to be occupied by newly evolved plants. Some of the rarest and most interesting plants occur in old landslides—the most recent habitats. The precipitous topography of scarp ridges and profound valleys presents many situations where colonising populations—not only from other mountains but also from the surrounding, vegetation communities—can be effectively isolated from mother populations, thus facilitating rapid evolution. Pleistocene climatic fluctuations causing repeated catastrophic extinctions, migrations and population expansions, and this rapid evolution act like an evolutionary pump, driving speciation that, in the plant world, is second to none.

CHAPTER

THE ULTIMATE ISLAND

'THE POSITION of Celebes [Sulawesi] is the most central in the Archipelago. Immediately to the north are the Philippine islands; on the west is Borneo; on the east are the Moluccan islands; and on the south is the Timor group: and it is on all sides so connected with these islands by its own satellites, by small islets and by coral reefs, that neither by inspection on the map nor by actual observation around its coast is it possible to determine accurately which should be grouped with it and which with the surrounding districts. Such being the case, we should naturally expect to find that the productions of this central island in some degree represented the richness and variety of the whole Archipelago, while we should not expect much individuality in a country so situated that it would seem as if it were pre-eminently fitted to receive stragglers and immigrants from all around.

As so often happens in nature, however, the fact turns out to be just the reverse of what we should have expected; and an examination of its animal productions shows Celebes to be at once the poorest in the number of its species and the most isolated in the character of its productions of the great islands in the Archipelago. With its attendant islets, it spreads over an extent of sea hardly inferior in length and breadth to that occupied by Borneo, while its actual land area is nearly double that of Java; yet its Mammalia and terrestrial birds number scarcely more than half the species found in the last-named island. Its position is such that it could receive immigrants from every side more readily than Java, yet in proportion to the species which inhabit it far fewer seem derived from other islands, while far more are altogether peculiar to it; and a considerable number of its animal forms are so remarkable, as to find no close allies in any other part of the world.'

Wallace was mystified. For sure, the depauperate nature of Sulawesi's fauna could easily be explained by its insular nature, where only animals which can survive an ocean crossing can make landfall. But its individuality, in the middle of a crowd, was something else again.

Charles Darwin had pointed out to Wallace that the depth of the sea was a measure of time. Shallow seas implied a recent connection: England, on a shallow shelf, was a mirror image of adjacent continental Europe. Borneo, Sumatra and Java, too, clearly lay on the same shallow bank or continental shelf. The corollary was that the unmeasurably deep trenches between Borneo and Sulawesi—in fact encircling Sulawesi—represented a long-continued separation; perhaps even a permanent separation. The idea was new and it was exciting.

Was Sulawesi the ultimate island? Reconstructions of around 160 million years ago certainly suggest that bits of western Sulawesi were already embedded in a drifting chunk of Gondwana (see Map 10, page 56). What a piece of magic must be an island adrift for 160 million years and 'peopled' through a chapter of accidents. The only other place in the world even remotely like Sulawesi is Madagascar. Madagascar separated from mainland Africa in the middle Jurassic, about 160 million years ago, and drifted south to reach its present position in the early Cretaceous, about 130 million years ago, well before mammals evolved. The late Cretaceous extinction event may well have wiped the Madagascan slate clean leaving its shores free to field for evolutionary survivors.

Madagascar's extraordinary mammalian fauna, evolving from a collection of ancient African faunal flotsam, reflects this long, independent history. Primitive offshoots from their respective orders, Madagascan Primates, for example, comprise three endemic families of lemur which presumably arrived on Madagascar's shores around 60 million years ago, shortly (geologically speaking) after the late Cretaceous extinction event; the Insectivora comprise an endemic family, the shrew-like tenrecs; the Rodentia include an endemic subfamily of murid rodents; and the Carnivora comprise a number of genera of viverrid (civet) that may all belong to a single endemic family. The only other land mammals that have reached the islands naturally are a pygmy hippopotamus that became extinct during the Pleistocene and a river-hog. Like Sulawesi, nearly all the native mammals of Madagascar are endemic, but only in Madagascar does the endemicity of most of these extend to family level.

The antiquity of Sulawesi's fauna is not so clear cut. Nor are its origins. Some animals, for instance the curious lemur-like tarsier *Tarsius spectrum* and the extraordinary pig-like Babirusa, seem to be more closely related to African animals than anything now living in Asia, let alone in the islands surrounding Sulawesi.

To account for this Wallace wrote: 'Celebes must be one of the oldest parts of the Archipelago. It probably dates from a period not only anterior to that when Borneo, Java and Sumatra were separated from the continent, but from that still more remote epoch when the land that now constitutes these islands had not risen above the ocean. Such an antiquity is necessary, to account for the number of animal forms it possesses, which show no relation to those of India or Australia, but rather with those of Africa; and we are led to speculate on the possibility of there having once existed a continent in the Indian Ocean which might serve as a bridge to connect these distant countries.'

Geographers at the time had already invented such a land and given it the name Lemuria, since it was to account for the puzzling distribution of lemurs. Wallace explained that while Lemurs have 'their metropolis in Madagascar, [they] are found also in Africa, in Ceylon, in the peninsula of India and in the Malay Archipelago as far as Celebes, which is their furthest eastern limit'. According to this idea, the isolated Indian Ocean Mascarene Islands and the Maldive coral group represented the left-over scraps of Lemuria.

Wallace was clearly grasping at straws.

The African connection and its long-standing association with South-East Asia is real enough though. For the 160 million years that it took for the bits of South-East Asia and India to get their act together, Africa hovered nearby, never far from Eurasia. But about 19 million years ago in the mid-Miocene, Africa finally began her approach in earnest.[13] Like a boat clumsily pulling alongside a jetty the collision was really a series of bumps. The first bump was between the Arabian and Turkish regions of

Africa and Eurasia respectively. While the continents grappled with one another, ancient carnivores, pigs, bovids and rodents scuttled across from Asia to Africa, while primates, elephants and creodonts (a group of ancient carnivores which gave rise to the whales and dolphins) passed from Africa to Asia. The continents then parted, only to briefly dock again around 16 million years ago when newer groups of primates, pigs and elephants, boarded Eurasia. The final docking at 12 million years ago saw the rhinoceros, hyaena and sabre-tooth cats disperse into Eurasia.

This final collision threw up the mountains of Arabia, Turkey and the Middle East and later the mountains of Spain and north-western Africa. Together with a global drop in sea levels, this buckling resulted in the evaporation of the Mediterranean, turning it into a salt lake of enormous proportions (making the expansive, salty Australian Lake Eyre look, in comparison, like a dried teardrop).

A Mediterranean heat-haze surrounded by desert mountains is a biogeographical barrier of continental proportions. It slashed across populations of animals found in both Africa and Asia, separating them. Primitive primates such as the lemurs and lorises, as well as monkeys, apes, porcupines, rhinoceroses, elephants and the pangolin, and birds, such as the hornbills, all found exclusively in Africa and Asia, are today represented by different genera in each of the continents.

Wallace called the Sulawesi tarsier *Tarsius spectrum* a 'curious lemur', clearly linking it to Africa. While tarsiers are primates, they are not lemurs. The four species of tarsier belong to their own family, the Tarsiidae. Sulawesi has two endemic species. The second species, *Tarsius pumilis* (which Wallace did not appear to be aware of), is only found in central Sulawesi. *Tarsius bancanus* is found on the other islands of the Sunda Shelf and *T. syrichta* in the Philippines.

Tarsiers are one of the world's smallest Primates with a head-and-body length of just 10 centimetres and a tail 20 centimetres long. They weigh about 100 grams.[79] They have such enormous eyes that they cannot rotate them in their sockets. Instead, it is the ball-like head which rotates—through a full 360 degrees. In their movements, tarsiers resemble tree frogs because both animals have elongated hindlimbs—the name 'tarsier' in fact refers to the elongated ankle or 'tarsal' region.[54]

In Sulawesi, the tarsier social unit comprises a monogamous adult pair and one or two immature offspring. The family all sleep together in a tree hole in the middle of a more or less permanent territory of about one hectare. This is defended from other tarsiers by song. Each morning just before the nocturnal family goes to rest they sing a complicated squeaking duet, the male initiating the song with a regular series of squeaks and the female joining in with a descending series of squeals which then rise in pitch to a fast climax. These squeaky operas apparently announce to the surrounding groups that the mated pair is present, alive and apparently happy. At sunset, the family wakes, interacts for 20 minutes or so, before leaping off in separate directions to hunt exclusively for insects.

Tarsiers are the only living decendents of an ancient group which was once widespread in Eurasia. Tucked away in island South-East Asia, they have survived their relatives by 30 million years.

Perhaps one of the most primitive animals on Sulawesi is the Babirusa, a name derived from the Indonesian babi (pig) and rusa (deer). The Babirusa is naturally found only on Sulawesi and nearby islands. Wallace explained that while this extraordinary barrel-bodied creature resembled a pig—he thought the wart-hog of Africa—its name was derived 'from its long and slender legs and curved tusks resembling horns'. The most remarkable feature of this remarkable animal is the development of the canines in the male. Wallace's description of them is vivid:

'The tusks of the lower jaw are very long and sharp, but the upper ones, instead of growing downward in the usual way, are completely reversed, growing upward out of bony sockets through the skin on each side of the snout, curving backward to near the eyes and in old animals often reaching eight or ten inches in length. It is difficult to understand what can be the use of these extraordinary horn-like teeth. Some of the old writers supposed that they served as hooks.'

Local wisdom indeed had it that the function of the curious canines was to allow the male to hook his heavy head on a branch while waiting for a female to pass by.[24] The real function of tusks, however, probably has more to do with tussles between males for females.

Taxonomists have placed the peculiar Babirusa within the suborder Suiformes which contains three living families: the Hippopotamidae (hippos), the Dicotylidae (peccaries) and the Suidae (pigs). Babirusa *Babyrousa babyrussa* is currently placed at subfamily level within the pig family. The concerns of Wallace, however, that 'the Babirusa stands completely isolated, having no resemblance to the pigs of any other part of the world' are still relevant today.

Morphologically, for instance, the Babirusa's snout is less specialised than in other pigs, in fact Babirusa does not use its snout to root like pigs but feeds on fallen fruit and grubs instead. Pigs have simple stomachs but the stomach of a Babirusa is more complicated. Its muscular structure, too, bears more of a resemblance to the peccarie, while its naked appearance gives it the cast of a hippopotamus. Its nearest relative, in fact, seems to be a 30-million-year-old northern hemisphere ancestor of the hippo. These 'Anthracotherids' have so far been found only in European fossil deposits. Hippopotamus evolved from an ancestral Anthracotherid around 11 million years ago. Fossils of a primitive hippopotamus, Merycopotamus, have been found in the 12-million-year-old Siwalik fossil site of India, and these indeed may be the ancestor of the Babirusa.[24] (There has also been an unsubstantiated discovery of an Anthracotherid skull in Timor dated at around 40 million years ago.[70]) If this finding proves to be accurate then Babirusa's relatives may have been very widespread indeed and oceanic mariners of extraordinary proportion!

One other native pig occurs on Sulawesi, *Sus celebensis*. It also is apparently a rather primitive representative of the *Sus* genus and may have evolved from old stock living during the Pliocene, around three to five million years ago, at which time it swam across to Sulawesi.

Also found in the Siwaliks, are fossils of an animal which may be an ancestor of another of Sulawesi's peculiar mammals, the Anoa *Bubalus depressicornis*.

In Wallace's time, the anoa (or the sapi-utan, literally forest cow) was an animal which caused much controversy, 'as to whether it should be classed as an ox, buffalo, or antelope. It is smaller than any other wild cattle and in many respects seems to approach some of the ox-like antelopes of Africa. It is found only in the mountains and is said never to inhabit places where there are deer. It is somewhat smaller than a small highland cow and has long straight horns, which are ringed at the base and slope backward over the neck'.

There are, in fact, two species of anoa, the Lowland Anoa *Bubalus depressicornis* and the Mountain Anoa *B. quarlesi*. Both are known for their ferocity, despite their relative diminutive size compared with their distant relative, the Asian Water Buffalo. The Mountain Anoa is only about 75 centimetres at the shoulder and the Lowland Anoa is around one metre. Sturdy looking animals, the Lowland Anoa has white forelegs and a longer tail than the Mountain Anoa which has uniformly coloured legs. Unlike the Water Buffalo, which is associated with wet grasslands and swamps, anoas are secretive animals of the forests.

The tarsiers, Babirusa and the anoas are all animals with a

long history in Eurasia. Where were the echoes of Australia that Wallace was so struck by? Apart from a handful of boisterous Australian-derived birds (in a situation resembling Lombok) the only Australian-derived mammals are two 'Eastern opossums' (cuscus). Wallace found these to be common, adding that they marked 'the furthest westward extension of this genus and of the Marsupial order'. He went so far as to suggest that Sulawesi may have once formed a western extension of a vast Australian or Pacific continent, although never having actually been joined to it.

Yet of the 63 terrestrial (non-flying) native mammal species in Sulawesi and its offshore islands that we know of today (in Wallace's time only 14 terrestrial species of mammals were known), only these two cuscuses are marsupials. And only these two are of clear Australian origin. In the light of this, there is no way that Sulawesi can be considered Australian.

Curiously, the only two native marsupials of Sulawesi are also the most primitive of the entire cuscus family, Phalangeridae. Restricted to Sulawesi and a few nearby islands, the Bear Cuscus *Ailurops ursinus*, in particular, is thought to have become isolated on Sulawesi earlier than 30 million years ago.[19] A striking and distinctive animal, the Bear Cuscus has short, coarse black fur variably tipped with yellow. It has a bear-like snout and is the only cuscus with round pupils—the other cuscus have ovoid or cat-like pupils—suggestive of day-time activity. With large feet and long limbs, it is a very un-possumlike possum. Indeed because so many of the features of this animal are so primitive it may be more appropriate to put it into its own subfamily, the Ailuropinae.[19]

The Small Sulawesi Cuscus *Strigocuscus celebensis*, a small plain-coloured animal which looks a little like the Australian brushtail possum, is also probably a relict group that has survived on its distant island home for millions of years. Noctural and occurring in pairs, this animal is a fruit-eater. A close relative, *Strigocuscus pelengensis*, exists only on Sulawesi's offshore islands of Peleng and Taliabu.

Wallace recorded only one civet in Sulawesi, the Common Malay Civet *Viverra tangalunga* which was no doubt introduced there by humans, who prized the musk-like 'civet' secreted by the animal's genital gland. Civets are part of a large family of Eurasian carnivores and Sulawesi does indeed have a native civet. One of the world's least-known carnivores, the Sulawesi Palm Civet *Macrogalidia musschenbroekii* is endemic to genus level and is the only carnivore native to Sulawesi. Very few Sulawesi Palm Civets have been seen, despite intensive searches, in the more than 100 years since it was first described. No wonder Wallace never got wind of it. The scant sightings indicate that, like most palm civets, the Sulawesi Palm Civet has a long, sinewy body with short legs and a long, banded tail. The head has a pointed, carnivorous snout. While the animal is a skilful climber and can hunt in the trees, it mostly skulks on the ground, hunting stealthily for small mammals and foraging for fruit.

The lack of carnivores is, of course, one of the hallmarks of an oceanic island. Even on the Sunda Shelf islands, Java, Borneo and Sumatra, carnivores are the first animals to go extinct when the sea level rises after a glacial peak. This is because carnivores are finely balanced at the pinnacle of a broad-based food pyramid. The density of prey on islands, particularly depauperate oceanic ones, is simply not great enough to support a self-sustaining population of carnivores.

Sulawesi's classic island nature is stamped on its fauna; if bats are removed from the calculation, the endemism of its mammalian fauna is complete at 100 per cent. (With the bats endemism is still a very high 62 per cent.) Even that most unique island in the world, Australia, does not have this level of endemicity. Ninety per cent of its non-flying mammals are endemic. In contrast, on the continent-like Sunda Shelf, 58 per cent of the non-flying mammals are endemic and the endemicity on each of the islands is lower still. Whatever else one believes about Sulawesian mammals, it becomes clear that in geologically recent times it has been extraordinarily difficult for mammals to get on or off this island.

Three-quarters of Sulawesi's mammals are made up of bats and murid rodents, both extremely mobile 'supertramps'. The rest of the mammalian fauna are composed of those mammals which, while not actually supertramps, are nevertheless inclined to disperse.[51] For instance, of the 16 families of marsupials and monotremes that exist in Australia, Sulawesi has two representatives. And these are from the mobile phalangers, the only marsupial family with a penchant for rafting.

From the west, Sulawesi has species from seven families out of 29 placentals (not including bats and murids) found on the Sunda Shelf. Six of these—tarsiers, pigs, buffalo, civets, macaques and squirrels—reach no further east than Sulawesi naturally.[51] Of the seventh family, the shrews, only three species of a family numbering hundreds of species occur east of Sulawesi, and these are widespread or known to have been carried by people.

As far as oceanic islands go, even this rather depauperate non-rat, non-bat faunal component is richer in species and genera than is usual. Partly this is because Sulawesi—an extensive land with great topographical relief—is in close proximity to a source area which is rich in species, genera and families. Partly it is because certain groups of animals have found Sulawesi to be an evolutionary Eden. Genetically revved-up, macaques, squirrels and shrews found Sulawesi to be an evolutionary race-track.

'*Cynopithecus nigrescens*' Wallace said referring to what we now know as *Macaca nigrescens*, 'is a curious baboon-like monkey, if not a true baboon, which abound all over Celebes and is found nowhere else but in one small island of Batchian [Bacan], into which it has probably been introduced accidentally. An allied species is found in the Philippines, but in no other island of the Archipelago is there any thing resembling them. These creatures are about the size of a spaniel, of jet-black colour and have the projecting dog-like muzzle and overhanging brows of the baboons. They have large red callosities and a short fleshy tail, scarcely an inch long and hardly visible. They go in large bands, living chiefly in the trees, but often descending on the ground and robbing gardens and orchards.'

Widespread, macaques are found throughout southern and eastern Asia including Taiwan and Japan and in extreme north-western Africa. Currently 19 species are recognised including seven taxa which are endemic to Sulawesi. Since some of these close relatives tend to hybridise when brought together, whether they are all true species or not is still a matter of dispute and a sure sign of recent dispersal.

The semi-terrestrial African ancestors of macaques originally moved to northern Africa about 10 million years ago. There, the drying of the Mediterranean catalysed the evolution of three different monkeys: two dependent on swamp and forest in a wetter northern Africa and one, the macaque, invading a variety of habitats in Eurasia. The oldest fossil macaque is dated at about 6 million years ago and is found in northern Egypt.[16] From here, one group moved north into Europe, flourished and then suddenly retreated, probably due to deteriorating conditions associated with the Pleistocene glaciations.

The other group moved east, quickly spreading around southern India, to Burma and Malaysia. It is thought that Sulawesi was colonised by ancestors from this wave, not more than two million years ago. *Macaca nemestrina*, found now on Sumatra and Borneo, is a close relative of these ancestors. Climatic fluctuations then drove the speciation of macaques in Asia and in Sulawesi from this original wave, forcing populations into often mountainous refuge areas during dry, glacial times and then allowing them to spread (as newly evolved species) during warm times. For instance, Pleistocene glaciations in Sulawesi have resulted in the speciation of the mountain-living *Macaca tonkeana* from the lowland *M. maura*. Today, the warm moist conditions have brought their—now reproductively incompatible—populations together.

The south-western peninsula of Sulawesi seems to be where landfall of the first macaques occurred. As if in confirmation, the macaque living on this peninsula, *Macaca maura*, is the most primitive of the Sulawesian macaques. *M. nigra*, living furthest away in Minahasa on the northern peninsula, is the most specialised, with the other taxa arrayed in between.

In contrast to the macaques, all six species of Sulawesian squirrels are endemic to generic level. The long period of isolation necessary to develop such individuality is reflected in the fact that, while the squirrels certainly had their ancestral beginnings in mainland Asia, their ancestors now no longer exist on the Sunda Shelf.[51] Like the Babirusa, tarsiers and Bear Cuscus, Sulawesi has proved to be an evolutionary outpost for squirrels.

The extraordinary Sulawesian Long-nosed Ground Squirrel *Hyosciurus heinrichi* is found only on the mossy floor of montane forests of central and northern Sulawesi.[51] *Rubrisciurus rubriventer*, on the other hand, is the largest squirrel on the island and occurs wherever there is suitable lowland forests, exploiting resources on the ground and lower storey. *Prosciurillus* is the third endemic squirrel genus, containing species of tree squirrels. These have neatly divided themselves according to size. *Prosciurillus murinus*, the smallest, forages in the lowlands; *P. leucomus*, the medium-sized one, lives in primary forest at all altitudes; and *P. abstrusus*, the largest of the three, is found only in montane forest—a sensible strategy given the high energy requirements of small mammals. The smaller the mammal, the larger is its surface area relative to size and the more difficulty there is in keeping warm.

Rules, of course, are meant to be broken, particularly, it

seems, in Sulawesi. Here, most of the eight species of Sulawesian shrew—some of the smallest mammals in the world—live from the lowland evergreen forests up to the mossy forests of the energy-sapping cold mountain tops. This fact becomes even more amazing when it is realised that shrews are extremely nervous, often literally dying of fright—even from thunder[54]—with heartbeats reaching 1200 times per minute. Tiny bundles of energy, they must eat their body weight each day and risk starving to death in a few hours. It is no surprise to discover that they live off the shrew equivalent of high energy bars: small insects, snails, millipedes, centipedes and spiders.

Predictably, all eight species of Sulawesian shrew are endemic at species level. These species are members of the widespread *Crocidura* genus which has 153 species distributed throughout northern Africa and Eurasia. The Sulawesian shrews show a confusing relationship, even though they are of the same genus. That they are Asian there is no doubt. Two of the species, *Crocidura rhoditis* and *C. nigriceps*, vaguely resemble the widespread Asian *C. fulginosa*, while *Crocidura levicula* resembles *C. monticola* from the Sunda Shelf. In a similar situation to Sulawesian squirrels, the other three Sulawesian shrews appear not to have any close relations in Asia. This outcome has at least two possible solutions. Either several ancestral shrews colonised Sulawesi over time, or all of the Sulawesian shrews evolved from the one ancestor. Either way it indicates a long, isolated history.

The great majority of mammalian species on oceanic islands are, however, murids. Sulawesi is no exception. Recent molecular studies[75] reveal a rapid evolution of murid rodents on Sulawesi beginning around six million years ago. Today the 36 known species, distributed among 14 genera, are all endemic, as are 10 of the genera. Some genera such as *Crunomys*, *Echiothrix*, *Tateomys* and *Melasmothrix* have no known living relatives on the Asian mainland, either having gone extinct or having evolved into different forms. Other genera such as

Margartamys, Lenomys, Eropeplus and *Haeromys* are related to some of the most primitive rats on the Sunda Shelf. Other genera again such as *Bunomys, Taeromys, Paruromys* and *Maxomys* have Asian relatives. That evolution continues apace in Sulawesi is indicated by the native *Rattus* on Sulawesi which is still actively radiating.

The endemic animals of Sulawesi—tarsiers, Babirusa, *Sus celebensis*, anoas, the Bear Cuscus, the Small Sulawesi Cuscus, all six species of Sulawesian squirrels, at least some Sulawesian shrews and most Sulawesian rodents—are all primitive families in their own right or primitive members of their families.

Why is Sulawesi a dumping ground for primitives? The answer lies in the dynamic nature of oceanic islands. One of the few studies done in the world on island dynamics—essentially the relationships between colonisation, extinction and speciation—has been carried out just north of Sulawesi, in the Philippine Archipelago.[27] The Philippines contain over 7000 islands and, with Indonesia, form the largest aggregation of islands in the world. Because they contain islands of widely differing sizes and history—some islands are strictly oceanic, some are fragments of once-larger islands and some had Pleistocene connections to the Asian mainland via Borneo—the Philippines are particularly appropriate as a testing ground for island biogeography hypotheses.

The Philippine Archipelago has a different geological and tectonic history to Indonesia, although both archipelagos were formed as a result of the collision of the Earth's tectonic plates. Most of Indonesia was formed as a result of the collision between the Asian and Australian plate. The Philippines, however, were thrust out of the sea as a result of the collision of two different plates: the Pacific tectonic plate and the South China Sea plate, an off-rider of the Asian plate which split off 32 million years ago (see Map 12, page 59).

The history of the Philippines began when the islands of the

Palawan arc, now fingering from north-western Borneo, rifted from Asia at around this time, as part of the South China Sea off-rider. At the same time, Luzon, now the largest island in the Philippines, began as a series of small, emergent volcanic islands more or less in the same position as it is today. Luzon continued to grow through volcanic activity at the boundary of the collision zone. Palawan emerged as an island in the Miocene, around 20 million years ago, slowly drifting to its current position.

The continued migration of the Pacific plate to the northwest from the Miocene onward caused the emergence of the other major islands of the Philippines (see Map 20 opposite) in the vicinity of Luzon, beginning with Mindoro around 10 million years ago. The large island of Mindanao, as well as Leyte and Samar emerged from the sea in their entirety probably around five million years ago. Parts of southern Mindanao, however, were caught up in earlier activity between the Australian and Asian plate, in the vicinity of Halmahera and may have emerged above the sea as long ago as 15–10 million years ago. Negros and Panay began to emerge around three million years ago and the Sulu islands, which string between Mindanao and Borneo, have been subaerial for only two million years or less.

Sea level drops during the Pleistocene caused the Philippine islands to merge into six major regions: Palawan, Mindanao, Luzon, Negros–Panay, Mindoro and the Sulu islands (see Map 20 opposite). From an experimental perspective, the configuration is ideal since it delimits regions of different ages and hence allows for comparisons of their fauna.

Therefore, starting with the youngest, the Sulu islands contain seven native species of which two, a shrew and a murid, are

---

Map 20. Paleogeographic reconstruction of the Philippines during Pleistocene times. (After Heaney, 1986.)

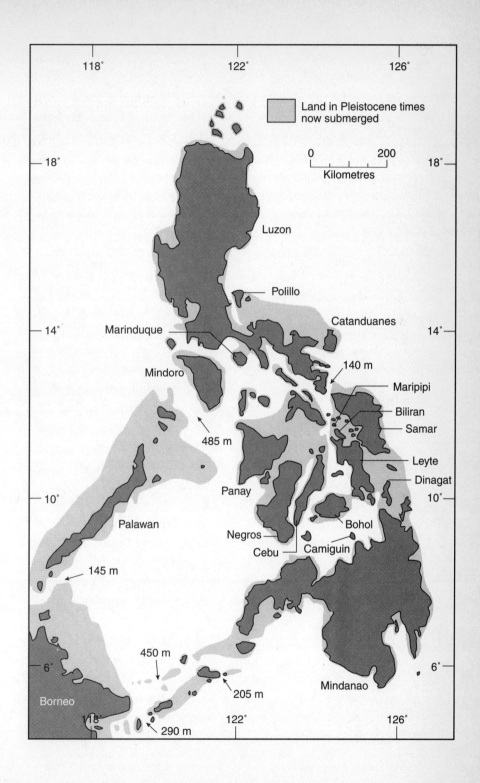

endemic. Of the islands in the Negros–Panay faunal region, only the island of Negros is well known. Its terrestrial mammal fauna consists of 13 species of which only three are certainly indigenous: an endemic shrew, a murid rodent and a deer. A tiger cat and the Long-tailed Macaque also occur in this region and while it is possible that these two animals are indigenous to the Negros–Panay region, both animals are also known to be transported by people.

The Mindanao faunal region has a total of 33 native terrestrial mammals; 79 per cent, or 26 species, are endemic. The endemism extends to four of the 21 genera. The high level of endemicity supports the fact that at least part of Mindanao is relatively old, perhaps older than Mindoro.

Alone among the different faunal regions, Mindoro's size and shape during the Pleistocene sea level yoyo has remained more or less unchanged due to the deep water surrounding the island. Of the 14 indigenous species of terrestrial mammals, 43 per cent (six) are endemic. All eight remaining species are shared with Luzon, the oldest oceanic island of the Philippines.

Much of the Luzon region remains unknown. Of the 29 indigenous species of terrestrial mammals known, 20 species or 69 per cent are endemic. These figures of endemicity may in fact be higher given that the Palm Civet *Paradoxurus hermaphroditus*, the Oriental Civet *Viverra tangalunga* and the Bearded Pig *Sus barbatus* may all have been introduced to the Philippines by people. Ignoring these animals raises endemicity in Luzon to 77 per cent, Mindoro to almost 55 per cent and Mindanao to 87 per cent.

The fauna of the Palawan islands, spread more or less equally between Borneo and the Philippines, has some surprises. Carnivores, pangolins, porcupines and mainland insectivores are not what anyone would expect of an oceanic island. Indeed these are virtually absent in the oceanic realm of the Philippines, just a few kilometres away. In addition, all Palawan's genera are shared with Borneo and only 11, or 44 per cent, of its 25 indigenous species are endemic. How is it that this island group, with an area not much larger than its near-neighbour Mindoro, can have such a drastically different mammalian fauna? (Mindoro naturally has a mammalian fauna of 11 species consisting mainly of rodents.)

If Wallace had been to Palawan he probably would have looked to sea depth to explain the dilemma. And he would have found what he suspected. Palawan is, in fact, an offshore island of the Asian mainland. Separated from Borneo by a channel only 145 metres deep, in the middle Pleistocene (when sea levels dropped by at least 160 metres), Palawan merged to become a narrow peninsula of the mainland. Palawan's relatively high species-level endemicity compared to the mainland is explained by its 160 000 year separation from Borneo since these very low sea levels. The depauperate fauna of the oceanic islands of the Philippines is all derived willy-nilly over water from the Asian mainland, probably via Borneo and Palawan.

Interestingly, while all the oceanic islands are impoverished compared to Palawan, the degree of impoverishment parallels the approximate age of the islands. Luzon and Mindanao being the least impoverished are also geologically the oldest. The Sulu islands, the youngest, are the most impoverished. While island area must certainly play a part in these figures, it is interesting to note that 10-million-year-old Mindoro, about three times smaller than the three-million-year-old Negros–Panay faunal region, has more than double the number of native mammals. In addition, the older, less impoverished faunas contain proportionately and absolutely more endemic species than the newer, more impoverished faunas.

Knowing the age and the approximate number of non-endemic native species of some of the younger islands, it is also possible to broadly estimate the rate of colonisation of oceanic islands by terrestrial mammals. The results are staggering.

Successful over-water colonisation, such as by rafting, only occurs around once every half-million years—vanishingly small—and this for islands which were separated by more speciose islands, during the late Pleistocene, by no more than 15 kilometres. In reality, successful colonisation must be even smaller than this, since most of the mammal species of the Philippines have actually evolved there. The roots of the family trees of the 40 known species of Philippine murids, for instance, represent a maximum of seven successful colonisation events. Looked at in this light, a successful over-water colonisation event is a virtual miracle.

Hence the near miraculous fauna of places like Madagascar and Sulawesi. These islands are like gambling time machines, sweeping up the faunal flotsam of aeons and allowing one event in a million to successfully spawn its own line. What are the chances that the next colonist to survive will be a competitor for precisely the same niche that has already been colonised? In habitat-rich Sulawesi the chances are smaller than breaking the bank. Hence different groups from different epochs have been able to survive side-by-side.

What of Sulawesi's antiquity? While parts of it may have been embedded in a drifting chunk of Gondwana since before 160 million years ago, whether or not any of it lay above the sea and for how long, remains unknown. What happens when we apply the results from the Philippines experiment—that the geological age of the environment is positively related to the degree of endemism—to Sulawesi?

Three mammal species hint at the antiquity of Sulawesi: the Bear Cuscus *Ailurops ursinus*, possibly endemic at subfamily level; the Babirusa *Babyrousa babyrussa*, currently placed at subfamily level within the pig family; and tarsiers, belonging to their own family, the Tarsiidae.

Tarsiers are fairly widespread in island South-East Asia and

THE ULTIMATE ISLAND

hence do not provide any real clues specifically for Sulawesi. Their nearest relatives appear to have been in Eurasia about 30 million years ago. The nearest known relative of Babirusa also seems to be a 30-million-year-old northern hemisphere 'Anthracotherid' ancestor. Anthracotherids are also found in Siwalik, the 12-million-year-old Indian fossil site. If, however, Babirusa is indeed a primitive pig then its relatives are likely to be any age from 30 million years to more recent times.

What about the Bear Cuscus, the most primitive of the entire cuscus family? Could it really have become isolated on Sulawesi earlier than 30 million years ago? Cicadas, those 'looking-glass' insects of Chapter three, which tend to mirror subaerial geological evolution, suggest no direct relationship with Australia. According to them, the Miocene-aged chunk of Australia which forms the eastern arc of Sulawesi (on which the Bear Cuscus is thought to have rafted from Australia) approached Sulawesi under the sea, being thrust above the sea as a direct result of the mid-Miocene collision with the apparently subaerial western arc, 20–15 million years ago. It was this collision that finally saw the spider-shaped Sulawesi of today emerge from the sea.

The cicadas' biogeographic relationships do, however, suggest an older western arc. But to reach this arc 30 million years ago, the Bear Cuscus would have to have been a good mariner indeed (see Map 13, page 60). The scant geological evidence of the western segment confirms only that it was land 10 million years ago.[2] Certainly Sulawesi is old, but there is no real evidence to suggest the great isolated antiquity of Madagascar or even of 45-million-year-old island Australia. Like an enigmatic woman, Sulawesi has, for the time being, veiled her true age.

And what of her relationship with Borneo? Does the rigid stand-off hide a discreet liaison? Certainly it seems true that the two have never been attached. Even during the lowest sea levels, perhaps some 200 metres lower than now, they kept their distance. It was during these times, however, that the few dispersing Asian mammals would have had the greatest chances of a successful landfall on Sulawesi. A closer look at the configuration of land and sea indicates that it is in fact a moot point as to whether the ancestors of Sulawesi's terrestrial mammals came from Borneo or any other place on the Asian mainland, since they were all joined at one time during the Pleistocene (see Map 18, page 87).

If most of the mammals rafted across, as seems probable, then the most likely place to 'board' a raft is on a river issuing into the sea. Today floating masses of logs and vegetation are found at the mouths of large, tropical rivers. As we have seen in Chapter five, one of the great rivers of ice-age Asia, with tributaries sweeping debris from Sumatra, Java and Borneo, would have debouched into the sea from the area between Java and Borneo, not far from the south-western peninsula of Sulawesi (see Map 18, page 87). The relationships and distribution of at least the Sulawesian macaques suggest this route as probable.

The Philippines certainly have received virtually all its terrestrial mammal fauna from Borneo via Palawan. What of the relationship between the Philippines and Sulawesi—by today's configuration as distant to one another as Borneo is to the Philippines? The Sangihe and Talaud islands stretch from the northern peninsula of Sulawesi toward Mindanao but even in the Pleistocene, sea barriers separated the two. Reflecting this, the mammals of Sulawesi and the Philippines show no close relationship with one another. Indeed, the northern extremity of the Wallace Line passes between the Philippines and Sulawesi, finally sweeping out to the Pacific.

In many respects this section of the Wallace Line is rather deceptive since the composition of both the Philippine and Sulawesian mammal fauna are both typical of oceanic islands; both vastly different from the composition of the mammalian mainland fauna. The differences between the Philippines and Sulawesi simply highlight the extraordinary degree of endemism inherent in relatively old, large, habitat-rich islands.

Indeed, other groups of animals such as birds and butterflies show that a significant relationship does in fact exists between the Philippines and Sulawesi (and hence between Sulawesi and Borneo!). The Philippines shares the overwhelming majority of its butterfly genera with the Malay Peninsula and the Greater Sunda islands; not surprising considering its close link via Palawan. Only a few genera are shared exclusively with Sulawesi.

These genera, however, comprise the significant portion of Sulawesian butterfly genera. At the species level, however, Sulawesi's strongest link is with the Moluccas. These patterns, which are also apparent in birds, reveal that the Philippines have played a vital role in the development of Sulawesi's more vagile biota; it was much easier for dispersing groups of butterflies and birds to filter across to the Philippines along the island chains linking it with Borneo. Once established on the Philippines these groups evolved rapidly under oceanic island conditions. Some naturally dispersed into Sulawesi where, once again, they radiated, naturally overflowing to Maluku.[38]

Like the islands of Nusa Tenggara (the Lesser Sundas) and Maluku (the Moluccas) the barrier between Borneo and Sulawesi is one between mainland biota and oceanic biota. But unlike these Johnny-come-lately archipelagos, Sulawesi is a mature, habitat-rich island which has nurtured its flotsam and jetsam colonists for tens of millions of years to produce an eccentric biota which is second to none.

Huxley's version of the Wallace Line, heading north with Borneo and Palawan on one side and the Philippines and Sulawesi on the other is, in fact, the more consistent line. Not that Wallace would have minded. To Wallace, mapping a boundary between the Asian and Australian faunas served multiple functions. It was a method for organising and simply communicating faunistic data, a potential device for predicting range limits of other species and, most importantly, a method of analysis that tested positively with his developing evolutionary hypothesis. In other words the Wallace Line was meant as a biogeographic tool. Indeed, Wallace himself was never hard and fast with the concept.

CHAPTER 9

THE BUTTERFLY EFFECT

The scale on which nature works is so vast—the numbers of individuals and the periods of time with which she deals approach so near to infinity—that any cause, however slight and however liable to be veiled and counteracted by accidental circumstances, must in the end produce its full legitimate results.

Alfred Russel Wallace. 'On the Tendency of Varieties to Depart Indefinitely From the Original Type', 1858.

DARWIN HAD sent a copy of his *Origin of Species* to Wallace, who wrote enthusiastic letters to relatives and friends. To Henry Walter Bates he wrote in 1860, 'never have such vast masses of widely scattered and hitherto utterly disconnected facts been combined into a system and brought to bear upon the establishment of such a grand and new and simple philosophy'.[43] To his brother-in-law, who quibbled over the heresy of the idea, Wallace wrote, 'it is the vast chaos of facts, which are explicable and fall into a beautiful order on the one theory, which are inexplicable and remain a chaos on the other'.[43]

Wallace's frequent-enough use of the word 'chaos' was, in a way, portentous. Wallace would have liked recent chaos theory. Despite its initially confusing title, chaos deals with non-linear, dynamic systems—like life. The fact that life rarely performs in some preordained, mathematically linear fashion would have been just common sense to Wallace. Indeed his notion that in the infinitely complex world of nature a small cause will result in a 'full' effect is not dissimilar to the classic imagery of chaos: that a butterfly stirring the air today in Peking will transform weather systems next month in New York.

Butterflies feature strongly in chaos theory. The classic pattern of chaos traces out a butterfly with two wings, a kind of double spiral in three dimensions. While the shape always stays within certain bounds it nevertheless signals pure disorder, since no point or pattern of points ever recur. Yet it also signals a new kind of order. And this was what Wallace was looking for. A theory in which the vast chaos of facts falls into a beautiful order. Just like his theory of evolution by natural selection.

In many ways the patterns on butterfly wings epitomise this new order. Just as the patterns of chaos are the patterns of nature, which everywhere exist as the infinite repetition of simple physical laws,[22] the apparent bewildering variety of butterfly wing patterns are in reality permutations of variations on simple themes. Even the most complicated wing patterns can be dissected into a small number of pattern elements. These elements form a system of themes that can be identified across thousands of species and are every bit as consistent in identifying butterfly species as bones are in identifying mammals.

This fact had also not gone unnoticed by Bates, who was instrumental in encouraging Wallace's lepidopteran addiction.

Still braving the butterfly-rich Amazon till 1860, Bates wrote, 'it may be said, therefore, that on these expanded membranes Nature writes, as on a tablet, the story of the modifications of species, so truly do all changes of the organisation register themselves thereon. Moreover, the same colour-patterns of the wings generally show, with great regularity, the degrees of blood-relationship of the species. As the laws of nature must be the same for all beings, the conclusions furnished by this group of insects must be applicable to the whole organic world'.[6]

The underlying order of the apparent chaotic variation of patterns of butterfly wings began to be revealed in the 1920s when two biologists named B.N. Schwanwitsch and F. Süffert independently invented an archetypal butterfly. This butterfly, which had wing patterns composed of fractal bands, circles or ocelli and a large fleck, called a discal spot, could produce much of the diversity of the world's butterfly wing patterns.

A further breakdown of the patterns of this archetypal butterfly revealed three developmentally identical systems of bands, known as symmetry systems (see Figure 2, page 162): a basal symmetry system, central symmetry system and border ocelli system. (The wing root band is only rarely present in butterflies.) It seemed that the butterflies had evolved a sort of 'do-it-yourself' developmental mechanism that produced easy-to-make patterns that could, nevertheless, generate a seemingly inexhaustible variety of individual textures and colours. And nothing, of course, transformed Wallace quite like such variety.

One day in south-western Sulawesi, for instance, on the way back to his tiny, two-room, bamboo hut, expansively named Mamajam, he captured a fine birdwinged butterfly. 'Trembling with excitement' he took this most beautiful of butterflies out of his net and found it to be in perfect condition. 'The ground colour of this superb insect was a rich, shining, bronzy black, the lower wings delicately grained with white and bordered by a row of large spots of the most brilliant satiny yellow. The body was

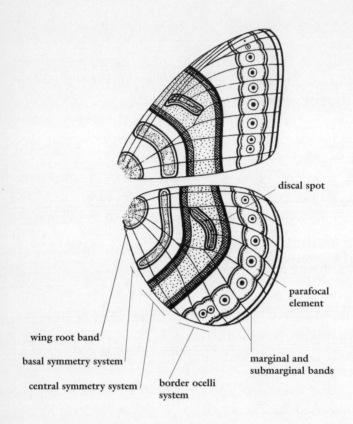

Fig. 2. Base plan of the butterfly wing pattern

Even the most complicated wing pattern can be broken down into a series of simple themes. Chaos incarnate, these simple themes can generate much of the dizzying diversity of the world's butterfly wing patterns. (Süffert, 1927 in Nijhout, 1991.)

marked with shaded spots of white, yellow and fiery orange, while the head and thorax were intense black. On the under side the lower wings were satiny white, with the marginal spots half black and half yellow.'

In Aru, Wallace had captured about 30 species of butterfly in a day. Most were rare and beautiful and hitherto only known by a few specimens from New Guinea. There he also caught 'the most magnificent insect the world contains, the birdwinged butterfly *Ornithoptera poseidon*. I trembled with excitement as I saw it coming majestically toward me and could hardly believe I had really succeeded in my stroke till I had taken it out of the net and was gazing, lost in admiration, at the velvet black and brilliant green of its wings, seven inches across; its golden body and crimson breast ... to feel it struggling between one's fingers and to gaze upon its fresh and living beauty, a bright gem shining out amidst the silent gloom of a dark and tangled forest'.

On one occasion when Wallace was rambling in the forest on the Kai islands in the South Moluccas, an old man with a keen sense of the absurd stopped to look at him. Wallace first pounced on an insect, then, as if carrying out a ritual, carefully stuck a pin through it and then delicately placed the neatly skewered exoskeleton in a little wooden box. The bemused old man stood very quietly till the end of the operation until 'he could contain himself no longer, but bent almost double and enjoyed a hearty roar of laughter'.

But catching butterflies was no laughing matter. To Wallace it was almost a matter of life and death and in Bacan in the North Moluccas, he nearly came undone. Here Wallace discovered a perfectly new and most magnificent species of birdwinged butterfly. It was one of the most gorgeously coloured butterflies he had ever seen. Its wings spanned 18 centimetres and were velvety black and fiery orange. 'The beauty and brilliancy of this insect are indescribable' he wrote, 'and none but the naturalist can understand the intense excitement I experienced when I at length captured it. On taking it out of my net and opening the glorious wings, my heart began to beat violently, the blood rushed to my head and I felt much more like fainting than I have done when in apprehension of immediate death. I had a headache the rest of the day, so great was the excitement produced by what will appear to most people a very inadequate cause.' Wallace named it *Ornithoptera croesus*.

The wings of a butterfly develop at an early stage, crumpled against the skin of the caterpillar—on the inside. Each surface of the wing is only a skin, one cell thick, so that the entire wing is only two cells thick. Veins branch through these thin wings in a pattern that is identical across most butterflies, dividing the wing into compartments. Colour diffuses from cell to cell within these compartments to form patterns.

To see how such a mechanism could generate the dizzying permutations of size, shape, colour and presence or absence of basic pattern elements, which make up the variety of butterfly wing patterns observed in nature, it is perhaps best to see the butterfly wing as a rather surreal landscape over which colour flows like water.

On this membranous landscape, imagine the colour issuing from lines, or fissures, on the wing cell midline, from the wing margins or at the wing veins. The colour flows across the gently undulating landscape in bands and pools in the regularly aligned valleys, or along the contours before it reaches the valleys. Some colour escapes to the groundwater underlying the surface topography. Occasionally a hill grows out of one of five points along the fissure lines, erupting colour like lava. The colour flows outward, down the cone of the volcano, spreading like a circle. Instead of mingling with the colour of the lowland, however, it reacts with it additively, each colour pushing the other into a different contour depending on the steepness of the slope and the shape of the volcano, which may not be perfectly conical.

Remarkably, computer modelling of combinations of any of five point sources and four line sources along the wing-cell midline and the margin, can indeed generate most of the diversity

THE BUTTERFLY EFFECT 165

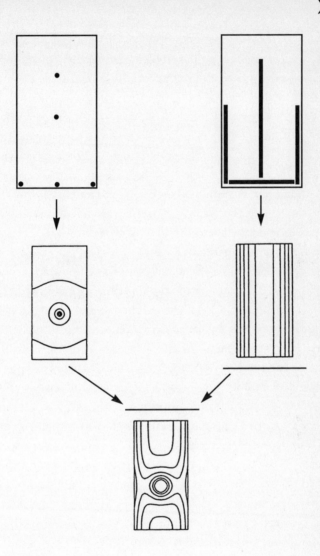

Fig. 3. Butterfly wingscape

These stylised butterfly wing cells show the sources from where colour seeps. Colour will not seep from all sources at once, but from one or two. For instance, when colour flows from the midline it forms a pattern of bands that will migrate to the edges. Colour oozing from the mid-source, say, will form a pattern similar to a bull's eye. If these patterns are then combined it will form the pattern indicated. Each contour represents a potential pattern. (After Nijhout, 1990 and 1991.)

of butterfly wing patterns observed in nature. In reality, the gradients or slopes are created by different concentrations of chemicals. At a certain concentration-threshold, the migrating colour pattern is locked into place along a gradient (see Figure 3, previous page).

Within the wing cells, it is actually the tiny scales that take on colours. When the caterpillar pupates, cells on the wing grow out into tiny, flattened scales. They partially overlap and through an electron microscope look like tiles on the roof of a house. Indeed the scales are mostly arranged in straight parallel rows perpendicular to the long axis of the wing, just as tiles are laid against the gable of a roof.

Since each scale has an all-or-nothing approach to taking on colour, the colour pattern is in effect a finely tiled mosaic of monochromatic scales. The colour-wash of the wing takes place one or two days before the emergence of the adult butterfly.

The major pigments in the wings of butterflies belong to four chemical types: melanin, ommochromes, pterins and flavonoids. Melanins—producing blacks and shades of brown—are by far the most common pigments and are responsible for most of the detailed patterns found among butterflies. Ommochromes are red-to-brown pigments. Pterins are white and yellow-to-red pigments. Flavonoids are common in plants and produce diverse and colourful pigments which include yellow, red and blue. Butterflies, like all other animals, are unable to synthesise flavonoids and obtain them from their food plants. White and yellow flavonoids are particularly widespread among butterflies. Scales that are going to become different colours also develop at different rates. In all species that have been examined so far, red and yellow are synthesised first and brown and black second.

The colour of a scale may be due to the presence of chemical pigments, or it may be a structural colour that comes about when light plays on the regularly spaced microarchitecture in the scale. All iridescent colours and almost all blues and greens are structural colours.

From an electron microscopic image each individual scale is a marvel; intricately fenestrated by parallel ridges bridged by tiny trusses. The diversity and complexity of fine structure within individual scales, causing iridescence, places butterfly scales among the most complicated extracellular structures manufactured by a single cell. Nano-small lattices, lamellae, ribs and laminations, use light as a tool to craft colour and texture. Satiny wings, for instance, are the result of an array of closely spaced microribs across the body of the scale; the colour blue is the result of light reflection from small tubular holes; and other iridescence is due to a lattice of evenly spaced rods (see Figure 4, page 168).

In the absence of pigment, scales will appear white because of the reflection of light from vanes and trapped air bubbles. Whether the white is chalky, pearly or satiny depends on the intensity of the scattered and reflected light and on the regularity of the scale's microarchitecture.

Most families of butterflies have evolved only one or two of the six or so different microarchitectured structures (see Figure 4, page 168). Within the single family of Papilionidae, the swallowtail butterflies, however, are found the whole diversity of structural colour found throughout all butterflies.

No wonder Wallace was so overwhelmed with the Papilionid butterflies. The elegant *Trogonoptera brookiana* (family Papilionidae) for instance, which Wallace discovered in Borneo, has scales with the ribs angled to form a partially overlapping array that act as multiple thin-film interference reflectors, producing the velvety black and metallic-green colours that gave him such joy.

Everywhere he came upon glorious butterflies which took his breath away and caused him to flounder for superlatives to describe them. In Timor, he netted the rare and beautiful

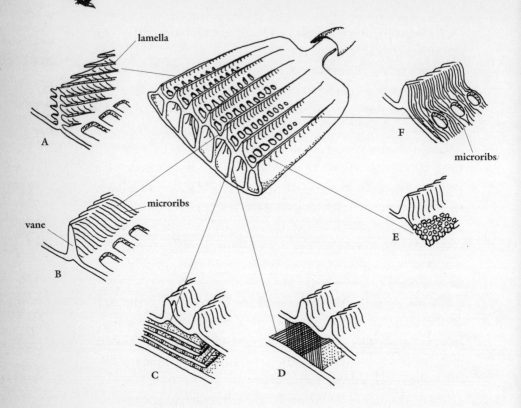

Fig. 4. Structural colour in butterfly scales

The diversity and complexity of the fine structure within butterfly wing scales places them among the most complicated extracellular structures manufactured by a single cell. Each of these six microarchitectural structures uses light as a tool to craft colour and texture. In 'A' parallel lamella forms a multiple thin-film interference reflector which plays with ultraviolet light. In 'B' an array of angled microribs produces iridescent colours, such as the blacks and greens of *Trogonoptera brookiana* that Wallace so admired. In 'C' the interior of the scale contains stacks of evenly spaced, sheetlike laminae which produce iridescent colours. In 'D' the interior of the scale is completely filled with a crystalline lattice which produces colours from diffracted light. In 'E' the perforations in the lamella scatters light and causes a pale blue colour. The microribs of 'F'—essentially a modification of 'B'—play with light to create a satiny appearance. (After Ghiradella, 1984, 1985 in Nijhout, 1991.)

Papilionid swallowtail butterflies, *Papilio aenomaus* and *P. liris*. In Ambon, at the heart of the Moluccas, the mythical Spice Isles, he gazed upon the brilliant blue, tailed wings of *Papilio ulysses*, 'one of the most tropical looking insects the naturalist can gaze upon'. Here, in the Moluccas, the insects were pre-eminently beautiful, even when compared with the varied and beautiful 'productions' of other parts of the Archipelago. Indeed only here is found the true birdwinged butterfly group—Wallace's heart-breakingly beautiful Ornithoptera—one of the most spectacular butterfly groups in the world.

Wallace had no explanation for these palpitatingly beautiful colours of the butterflies. Without electronic microscopes and modern techniques, he could not possibly envisage the developmental process that gave rise to it, nor the genetic programming that fundamentally controlled it. He had never heard of genes.

Ironically, at the same time that Wallace saw evolution reflected in the beauty of a tropical butterfly's wing, an Augustinian monk on the other side of the world saw it reflected in the colours of the common garden pea. Gregor Mendel (1822–1884) at the age of 21 joined an Augustinian monastery in Austria where he was to spend the rest of his life. With a background in palaeontology, physics, mathematics, chemistry and botany from the University of Vienna, he was a trained scientist. In 1854, he began some rigorous breeding experiments with the garden pea to discover a solution to the mystery of the evolution of life. In a skilfully designed experiment he analysed what would happen if he bred several different pure strains of garden pea—for example, peas which produced only purple flowers, peas which bred true with white flowers, peas which were wrinkled and peas which were round—and then mixed them.

In Mendel's time people thought about heredity in terms of essences, a little like blending coffee: add this essence to that and the outcome will be a blending of both. Eggs and sperm were thought to have essences of hair, arm, head and so on. A new individual would be a blending of both parents. Mendel showed that this was wrong. He saw no blending of characters: white and purple flowers did not produce a mauve offspring. Instead he showed that inheritance results from a few very important particles (now called genes), which direct the synthesis of new individuals.

Evolution, of course, is fundamentally a genetic process. Within the wing cell, different concentrations of diffusable chemicals trigger genes which control the injection of other colours. The entire process of butterfly wing modelling is under genetic control. How a colour reacts with other molecules, how it diffuses through the cell, where it pools and the concentration of its source are regulated by enzymes which are coded for by genes.

Generally, the genetic toolbox models the modular structure of wing patterns, with one level of genes affecting the theme pattern across the wing and one group affecting the same pattern, but independently, within each wing cell. This often means that the bands and circles of the symmetry system don't quite line up. In addition, most genes affect the pattern on only one wing surface: either forewing or hindwing; and either upper or lower surface. Some genes affect only certain elements of a pattern within a wing cell, say just the central symmetry pattern element, or the border ocelli element. There are also genes which affect colour and other genes which finetune the various pattern features. Ultimately genes program the randomness, the chaos, the variety of form and colour, that natural selection moulds into the rich tapestry of life.

Natural selection works at the species level and provides the bounds within which butterfly species vary. In this way the patterns on a butterfly's wings are images of a slice in time. They represent a frozen moment in evolution.

One of the most extraordinary of these images is the remarkable dead-leaf butterflies of the genus *Kallima*. When settled, the butterfly shape-shifts into a leaf. Wallace, in Sumatra, was astonished when he spied it among the dry woods and thickets. While he saw it commonly enough, it mostly eluded him by simply alighting amongst the vegetation and dissolving into the background, so complete and marvellous were the fine details of the butterfly's disguise. In amazement Wallace wrote:

'The end of the upper wings terminate in a fine point, just as the leaves of many tropical shrubs and trees are pointed, while the lower wings are more obtuse and are lengthened out into a short, thick tail. Between these two points there runs a dark curved line exactly representing the midrib of a leaf and from this radiate to each side a few oblique marks which well imitate the lateral veins. These marks are more clearly seen on the outer portion of the base of the wings and on the inner side toward the middle and apex and they are produced by striae and markings which are very common in allied species, but which are here modified and strengthened so as to imitate more exactly the venation of the leaf. The tint of the undersurface varies much, but it is always some ashy brown or reddish colour, which matches with those of dead leaves. The habit of the insect is to rest on a twig and among dead leaves and in this position, with the wings closely pressed together, their outline is exactly that of a moderately sized leaf slightly curved or shrivelled. The tail of the hindwings forms a perfect stalk and touches the stick while the insect is supported by the middle pair of legs, which are not noticed amongst the twigs and fibres that surround it. The head and antennae are drawn back between the wings so as to be quite concealed and there is a little notch hollowed out at the very base of the wings, which allows the head to be retracted sufficiently.'

Figure 5 on page 172, shows how the elements of the *Kallima* could be easily derived from the patterns of an archetypal butterfly. Fused bands of the basal symmetry system and the central symmetry system have become aligned to form

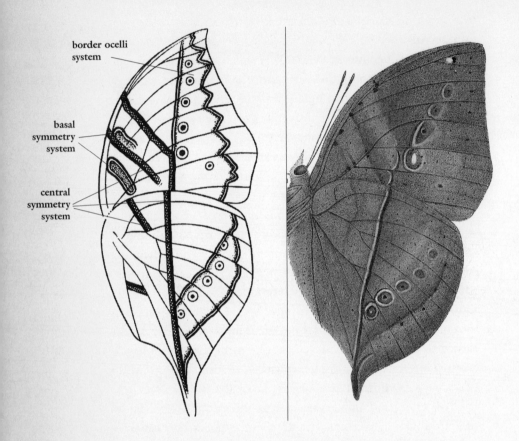

Fig. 5. Derivation of the wing pattern of *Kallima inachus*

The wing patterns of the remarkable dead-leaf butterfly of the genus *Kallima* can easily be derived from the patterns of an archetypal butterfly. (Archetype by Süffert, 1927 in Nijhout, 1991; and hand coloured lithograph of male *Kallima spiridion* from H. Grose-Smith and W.F. Kirby's *Rhopalocera exotica*, Vol. 2, 1892–1897, courtesy The Australian Museum, Sydney.)

single lines mimicking the veins of the leaf. Certain pattern elements in some wing cell compartments have been altogether lost, to help with the ruse. The large midrib pattern of the 'leaf' consists of, initially, the outermost band of the central symmetry system. Near the wing tip, however, this band turns and becomes a 'side vein' and the inner row of the border ocelli take over the disguise of the 'midrib'. To Wallace, the mimicry was further enhanced by the presence of small black dots, so closely resembling the way in which fungus grows on leaves, that it seemed as if the fungi had grown on the butterflies themselves.

Wallace noted that the butterfly had a strong and swift flight—enough to escape any predators. Accordingly, the upper wing of the butterfly was not disguised and appeared a conspicuous rich purple, variously tinged with ash and across the forewings a bar of deep orange.

The pattern on the top and bottom of butterflies wings are, in fact, frequently different. As butterflies rest with their wings closed, the bottom of the wing is exposed to the environment. A filigree of dark and light earth colours is used to produce a general purpose pattern that can blend into the colours of bark, leaf litter or soil—the places where butterflies alight. A close look at the details of the pattern will reveal that it is nevertheless ringed or banded, simply because it is developmentally easy for a butterfly to make patterns that way. Evolution, however, has selected the general texture of the lower wing rather than the pattern details. In complete contrast the upper wing pattern functions largely as a signalling device so that its patterns are bold and uncluttered with background noise.

Had Wallace taken heed of Mendel's work, he may have been able to make sense of one of the most bizarre natural history observations he ever witnessed during his adventures in the Malay Archipelago.

In 1861, in Sumatra, Wallace came upon a not unfamiliar butterfly, *Papilio memnon*, a splendid butterfly of which the male is a deep black colour, dotted over with lines and groups of scales of a clear ashy blue. Its wings are 12 centimetres in length and it has rounded hindwings. The females are coloured differently and some females are mimics of other rather nasty tasting butterflies. In nature there are at least 26 different forms of *P. memnon*. Each of these females can produce a brood which contains, as well as males and male-like females, one or more different female forms which mimic other butterflies.

Wallace had never seen anything like it. Incredulously, he likened the phenomenon to an Englishman on a remote island who had two wives (hence the remote island), one a black-haired and red-skinned Indian, the other a woolly headed, sooty skinned negress, with each mother 'capable not only of producing male offspring like the father and female like herself, but also other females like her fellow-wife and altogether differing from herself'.

Mimicry is, in fact, a common adaptation among the butterflies and *Papilio memnon* in particular is known for the spectacular polymorphism—the occurrence of varied forms within a species—that has evolved as different geographic races have come to mimic an array of different bad-tasting species in their respective regions.

The major differences in the forms of *P. memnon* are due to a cluster of five genes at the same locality on chromosome–X, the female chromosome. One gene determines whether or not a female butterfly will have a tail; another determines the colour patterns on the forewing; another, the colour of the small region near the base of the forewing called the epaulette; another, the colour of the hindwing; and the last gene, the colour of the body. Each gene can produce two or more different outcomes. The gene that affects the tail, for instance, has two possible outcomes: the presence or absence of the tail. On the other hand there are, so far, three known outcomes for body colour and four for the colour of the hindwing. Geneticists call these outcomes alleles, that is, there are two alleles for the gene determining tails and four for the gene determining the colour of the hindwing. If these genes acted independently it would be a little like rolling dice and there would be around 100 different outcomes or forms of *P. memnon*.

The five genes, however, do not act independently. Instead, the cluster acts as a 'supergene' so that certain outcomes are tightly linked with others (so, for argument's sake, a scarlet epaulette will always have a tail). The rationale for the evolution of the supergene is that they provide a mechanism for stabilising outcomes. In other words the dice are loaded so that only favourable outcomes occur; in the case of *Papilio memnon*, so that females that mimic other bad-tasting butterflies are produced. Natural selection has ensured that these females—and their supergenetic combination—have a place in history.

At first glance, butterfly wing patterns are a kaleidoscope of complexity, mirroring the infinite variety of life that left Wallace astonished. Even the most complicated wing pattern, however, can be broken down into a series of simple themes, generated by the repetition of the genetic process. In this way, butterfly wing patterns are essentially snapshots of chaos, with natural selection riding the boundary.

CHAPTER 10

THE RED APE OF ASIA

> Let us ask the hypothetical and simple question: 'What have Orang-utans done to us that we have driven them to near extinction?' The answer to this is easy (as indeed for most other species in this predicament) and there would be few who could quarrel with the answer: nothing, absolutely nothing.
>
> *Gisela Kaplan and Lesley Rogers.* Orang-utans in Borneo, *1994.*

THE 'ROAD' consisted solely of slippery tree trunks laid end-to-end. Along this, the Dayaks sped, their bare, broad native feet finding purchase easily, despite the heavy boxes they were burdened with. Alfred Russel Wallace, boot-shod, slipped and stumbled along behind, occasionally sinking ankle-deep into the swamp which extended 30 kilometres to the coast in a flat jungle-covered plain, punctuated here and there by a few isolated hills and mountains. A long, narrow river wound through the plain, gleaming gold through the deep green of the Bornean dipterocarp forest.

The Dayak-made road was over-arched by tall trees and led to a clearing at the base of one of the mountains. When he was not focused on his blundering boots, Wallace's head was craned toward the lofty forest overhead. Occasionally he glanced admiringly at the agile Dayaks, who were bearing, with ease, all his worldly goods, including his heavy boxes of specimens. He noticed a few birds and insects and some very handsome orchids in flower. But his chief purpose in coming to stay here in March, 1855, was to see the great man-like ape of Borneo, the Orang-utan.

In the clearing, several rude houses had been erected. Here lived a number of Chinese and Dayak workmen who had been clearing the jungle for a railroad which would transport newly discovered coal to the river. The place was a mess, with rotting wood lying about everywhere from the clear-felling operations. Wallace was delighted. He knew that in the midst of virgin forest this amount of decaying material was ideal for collecting beetles and other insects. And, to boot, the location in southern Sarawak, on the border with Indonesian Kalimantan, was in the heart of Orang-utan-inhabited forest. With excitement, he craned his head this way and that, hoping to spot a 'mias', as the Dayaks called the Orang-utan.

Wallace did not have to wait long. A week after his arrival at the coal-works he saw his first Orang-utan. Some rustling in the trees distracted him from insect collecting. Looking up he saw the red-haired ape immediately, its backlit coat a fiery orange. It was moving along slowly, minding its own business, hanging from the branches by its arms and swinging from tree to tree. Wallace watched in awe till it was lost in the jungle.

A fortnight later he heard another Orang-utan feeding in a tree in a swamp just below his newly erected, two-room hut. He grabbed his gun and hurried out. After the second shot, the young half-grown male Orang-utan fell heavily to the ground, dead. This was the first of 17 Orang-utans that Wallace shot.

The animals, typically, were difficult to kill. One day, he had just arrived at his hut after a session of collecting beetles, mostly longicorns and other wood-eaters, and also butterflies and other insects. He knew he had at least 24 new species and was carefully examining them when his young assistant Charles Allen rushed into the hut, out of breath and full of excitement. A huge Orang-utan had been seen near the coal mine. It had come down the mountain and was trying to find a way around the cleared area to the low, swampy ground. Quickly they grabbed guns and ammunition and set out to bag the giant.

Wallace's heart was racing when he saw the great, red, hairy body, with a huge black face gazing down from the heights, as if wanting to know what was making such a disturbance below. Wallace fired into the black face, but the Orang-utan made off, moving rapidly and noiselessly for one so large. Big, angular fragments of rock from the mountain above, and a jungle thick with hanging and twisting creepers, slowed the men. The cleared area around the coal mine slowed the Orang-utan and as it halted, Wallace pumped four shots into it. With one leg hanging down uselessly, the Orang-utan climbed to the loftiest tree in the area where it wedged itself into a fork, hidden by thick foliage.

Wallace was afraid the Orang-utan would remain and die in this position and he would lose his specimen. 'As it was nearly evening,' he wrote, 'I could not have got the tree cut down that day. I therefore fired again and he then moved off and going up the hill, was obliged to get on to some lower trees, on the branches of one of which he fixed himself in such a position that he could not fall and lay all in a heap as if dead or dying.

... One of the Dayaks took courage and climbed toward him, but the mias did not wait for him to get near, moving off to another tree, where he got on to a dense mass of branches and creepers which almost completely hid him from view. The tree was luckily a small one, so, when the axes came, we soon had it cut through: but it was so held up by jungle ropes and climbers to adjoining trees that it only fell into a sloping position. The mias did not move and I began to fear that after all we should not get him, as it was near evening and half a dozen more trees would have to be cut down before the one he was on would fall. As a last resource, we all began pulling at the creepers, which shook the tree very much and, after a few minutes, when we had almost given up all hopes, down he came with a crash and a thud like the fall of a giant.'

With astonishment Wallace noted that the Orang-utan measured 'seven feet three inches' across his outstretched arms and from the top of his head to his heel, 'four feet two inches'. Its injuries were severe. Both legs were broken, one hip-joint and the root of the spine completely shattered and two bullets were found flattened in his neck and jaws. Yet the Orang-utan was still alive when he fell. Wallace was occupied the whole of the next day, preparing the skin and boiling the flesh off the bones to make a perfect skeleton, which Wallace sent to the Museum at Derby, England.

What can we say about this mild-mannered, kind and thoughtful man, one of the most forward thinkers of the day, except that he was still very much a product of his time. It is probably true that Wallace did not feel the slightest twinge of pity when he aimed at and shot Orang-utans, even when they were looking directly at him from a tree, peacefully feeding. He was after all, a nineteenth century naturalist-adventurer and the 'Far East' was, after all, an exotic goal in which to shoot whatever seemed of interest. (And to be fair, modern day museum scientists are little different.)

That there was much interest is indicated in Wallace's 'Eastern Collections', which amounted to 310 mammals (including the 17 Orang-utans), 100 reptiles, 8050 birds, 7500 shells, 13 100 lepidoptera (butterflies), 83 200 coleoptera (beetles) and 13 400 'other insects'—altogether 125 660 specimens of natural history.

In Borneo, Wallace also shot the curious Flying Lemur *Cynocephalus variegatus,* which he said 'possesses such a remarkable tenacity of life that it is exceedingly difficult to kill by ordinary means'. It was this 'tenacity of life' that led him to compare the Flying Lemur with a similarly hard-to-kill marsupial cuscus. Both animals, Wallace said, had a small brain. In fact he thought the Flying Lemur formed a transition with the marsupials since the young of both were born blind and naked.

Today taxonomists still have a problem in placing the Flying Lemur since its nearest relative appears to be a 60-million-year-old group which lived in North America. A curious animal with a unique, primitive skull and, in females, mammae located almost in the armpits, Flying Lemurs have large, gliding membranes attached between the neck and the sides of the body, incorporating the fingers, toes and tail. This membrane is larger than in other gliding mammals whose membranes are stretched only between the limbs, leaving the toes and tail free. So different is this animal that an entire order, Dermoptera, has been erected for it.

Classificatory compromises have also been made for the Oriental tree shrews or tupaias and the scaly anteaters or pangolins. For many years, the 11 different species of tupaia, which are endemic to the Oriental region, were regarded as very primitive primates. Then in 1974 it was determined that they were more likely to be primitive insectivores.[54] This, in turn, was refuted and finally a new order, Scandentia, was erected for the them. Wallace felt that they closely resembled squirrels and indeed the Indonesian word for squirrel is 'tupaia', a name which has now been hijacked by science to label the tree shrew family, Tupaiidae, and the genus *Tupaia*.

Seven species of pangolins are distributed throughout Asia and Africa, of which only *Manis javanica* extends to all the Sunda Shelf islands of Sumatra, Borneo and Java. Pangolins have a diet of termites and ants which they sweep up with their tongues, amazing organs which have muscular roots that pass down through the chest cavity and anchor to the pelvis. Originally taxonomists had lumped pangolins with anteaters, sloths, armadillos and the African aardvark. Later, however, it was assumed that the rainforest-loving pangolins' adaptations were simply the same solutions to a similar way of life (coevolution) and a new order, Pholidota, has been erected for them.

These curiosities are just a few of the many mammals of the Sunda Shelf forests of Sumatra, Java and Borneo. In the complex three-dimensional structure of these continental rainforests animal habitats are as layered as the vegetation. No wonder Wallace thought it a different world to that of the oceanic islands. The Sunda Shelf rainforests also show regular patterns in the composition of their mammal faunas—for example, Sunda Shelf mammalian faunas have about 10 per cent of their fauna made up of Primates, whether the island has 130 species (Borneo) or only eight (Redang Island, off the eastern coast of Malaysia). Insectivores such as tree shrews and ungulates such as deer, pig and mousedeer also make up a small, but consistent proportion of the fauna, as do the Dermoptera (Flying Lemurs) and Pholidota (pangolins).

And in what could not be a starker contrast to the oceanic island realm, every niche is filled. Terrestrial animals creep on the forest floor: tiny mousedeer or chevrotains with delicate legs as thin as pencils; diminutive muntjacs or barking deer, the stags of which bark fiercely to keep other males away and attract females; the black and white Malay Tapir which becomes invisible at night in the forest when moonlight on vegetation assumes the same colour; the Sumatran and Javan rhinoceroses, both on the verge of extinction.

THE RED APE OF ASIA

Arboreal creatures move acrobatically through the canopy, and a wide range of animals creep, crawl, leap, swing, scuttle and glide in-between. Monkeys, Wallace observed, were one of the most characteristic faunal features of the region. All sorts of Primates, including a Slow Loris, tarsiers, two species of macaque, eight species of leaf monkey and six different species of gibbons (close relatives of the Orang-utan), confirm this. In addition, at least 30 types of squirrel range from the canopy to the forest floor during the daylight hours; while the night draws around 13 different types of flying squirrels from their holes.

In contrast to the oceanic islands, where carnivores are virtually absent, at least 29 different carnivores exist in the Sunda Shelf islands. In Wallace's time the tiger was one of the most feared.

'The island of Singapore consists of a multitude of small hills, three or four hundred feet high, the summits of many of which are still covered with virgin forest. The mission-house at Bukit-tima was surrounded by several of these wood-topped hills, which were much frequented by wood-cutters and sawyers and offered me an excellent collecting ground for insects. Here and there, too, were tiger-pits carefully covered over with sticks and leaves and so well concealed, that in several cases I had a narrow escape from falling into them. They are shaped like an iron furnace, wider at the bottom than the top and are perhaps fifteen or twenty feet deep, so that it would be almost impossible for a person unassisted to get out of one. Formerly a sharp stake was stuck erect in the bottom; but after an unfortunate traveller had been killed by falling on one, its use was forbidden. There are always a few tigers roaming about Singapore and they kill on an average a Chinaman every day, principally those who work in the gambir plantations, which are always made in newly cleared jungle. We heard a tiger roar once or twice in the evening and it was rather nervous work hunting for insects among the fallen trunks and old sawpits, when one of these savage animals might be lurking close by, waiting an opportunity to spring upon us.'

Today the Java and Bali subspecies of tiger are extinct and only about 500 individuals are thought to exist of the Sumatra Tiger.[54] Singapore has no wild tigers. The leopards of Sumatra and the Clouded Leopard of Sumatra and Borneo suffered a similar plight, where only a few remain.

Other wild cats which hunt in the Sunda Shelf rainforests include the Marbled Cat, the Bornean Bay Cat, the Asian Golden Cat, the Leopard Cat, the Fishing Cat and the Flat-headed Cat. Further carnivores include the dhole or wild dog, now very rare in Java and Sumatra, the omnivorous Malayan Sun Bear, a number of weasels, martens, ferret badgers, river otters, clawless otters, civets, Oriental linsangs, palm civets, otter civets and mongooses.

As fascinating as these creatures are, Wallace paid scant attention to them, probably because they were already well known in his time and would not bring the high prices of the little-known Orang-utan specimens.

In 1960, researcher Barbara Harrisson was appalled by the state of neglect and sheer lack of interest in the Orang-utan. Unbelievably, the wild Orang-utan had remained entirely unstudied. 'There is no reliable, first-hand, scientific information whatever on daily journeys, arboreal/terrestrial movement, natural groups, food gathering, effect on vegetation habit, calls and sounds.'[34] This was despite the fact that over 900 Orang-utans were held in zoos worldwide. Indeed, only in the last 20 years has any continuity of information about Orang-utans in their natural state been obtained. The 'forgotten ape' had an image problem.

The great apes of the world include the Gorilla *Gorilla gorilla*, the Orang-utan *Pongo pygmaeus*, the Common Chimpanzee *Pan troglodytes*, the Pygmy Chimpanzee or bonobo *Pan paniscus* and Humans *Homo sapiens*. All except Humans are now endangered. In the early 1920s the Orang-utan was considered the least interesting of apes since it appeared to lack the intelligence of the Chimpanzee and the brutality of the Gorilla which had an unbeatable image:

> The ape was a great bull, weighing probably three hundred pounds. His nasty, close-set eyes gleamed hatred from beneath his shaggy brows, while his great canine fangs were bared in a horrible snarl as he paused a moment before his prey.
>
> *Edward Rice Burroughs.* Tarzan of the Apes, *1912.*

Surprising then, that the frontispiece of Wallace's *The Malay Archipelago* depicts an enraged male Orang-utan with its teeth sunk into a Dayak hunter's upper arm and one of its long, powerful arms wrenching the man's spear from him. In the background, four Dayaks are springing to the rescue with clubs, choppers and spears. In reality an enraged Orang-utan is a rare sight indeed.

A few minutes before the moment illustrated, the Orang-utan had been feeding innocently on the young shoots of a palm by the riverside near to a Dayak house. A number of Dayaks ran out to intercept him. Alarmed, the Orang-utan tried to escape to the jungle but one of the men tried to run a spear through the Orang-utan's body. In self-defence, the animal took the spear in his hands and 'got hold of the man's arm, which he seized in his mouth, making his teeth meet in the flesh above the elbow, which he tore and lacerated in a dreadful manner'. The man was rescued by the others who quickly destroyed the Orang-utan with their spears and choppers.

Orang-utans are, in fact, the most placid and gentle of the apes as Wallace found when he nursed an infant Orang-utan he found embedded face-down in the swampy ground after he had killed the mother. The baby 'was about a foot long' and had evidently been hanging onto its mother when she first fell. It began to cry after the mud was cleaned out of its mouth and on spying the long, scruffy hair hanging from Wallace's face—as long and unkempt as its mother's red hair—it latched onto the beard with long fingers which bent inwards to form complete hooks. When Wallace carried the little animal it was very quiet and content but when he laid it down by itself it would invariably cry. Wallace lost several nights sleep.

The behaviour of the baby Orang-utan can be understood in the light of recent studies of Orang-utan mother–infant interactions.[34] Wallace's adopted infant was used to being fondled and touched, its mother spending hours investigating the infant's arms and legs, occasionally taking the small hands gently into hers and inspecting the fingers. After the first week of mollycoddled life, the infant would have graduated to clinging to the mother's long, red hair. Indeed, for the first six months of life an Orang-utan infant does not leave the mother's body at all, but crawls all over it and may hang from it. The mother's single-minded care-giving is the longest and most intense of any Primate except Humans' and lasts around five years.

No wonder Wallace observed his baby Orang-utan, not more than a few days old, clinging desperately with all four hands to whatever it could dimly focus its newborn eyes on, including his long, Orang-utan-like beard. As a last resort, when left deprived of anything warm and hairy, the infant crossed its arms, hugging itself and poignantly grasping its own long hair that grew longest just below the shoulder.

Wallace's little pet suffered terrible diarrhoea and fever and this finally killed it after three months. Today it is well known that Orang-utans suffer the same diseases as humans. The infant had put no weight on since it was first captured. A diet of ricewater, rice and biscuits was a poor substitute for mother's milk and the 400 or so other jungle foods that Orang-utans are known to eat, including fruit, leaves, shoots, insects, soil and bark. After the infant's death, Wallace pragmatically preserved its skin and skeleton. It was then that Wallace noticed that the infant had broken an arm and a leg, no doubt during its fall. These had mended irregularly as crooked swellings in the small skeleton.

Orang-utans are solitary animals, known to mind their own business. Males, in particular, show a marked aversion to each other. Indeed, they are like cranky bachelors who like to be alone for at least 90 per cent of the time. But, like most men, a female will occasionally catch a male's eye and he may choose to consort with her—but not for more than a few weeks at a time. Females, too, are solitary, but not often actually 'alone' since they spend most of their lives with offspring, which they raise without help. Subadult females are the most social, spending most of their time with either their mother and younger offspring or with other subadult females and subadult males. Subadult males will choose the company of females over males. Generally, however, there can be any number of groupings, friendships or brief associations and up to 67 different associations have been reported.[34] Apart from a strong aversion between adult males and a general preference for a solitary lifestyle, anything goes for this laid-back ape.

Anything, that is, except disturbance by humans. Wallace was fascinated by the fact that the Orang-utans often howled at him in rage, in a strange voice, he said, like a cough, breaking branches with their hands and throwing them down. Orang-utans, not being biased, also carry out this behaviour on modern-day researchers even though they no longer come to kill. Gisela Kaplan and Lesley Rogers, who study Orang-utans in Sabah, regard this behaviour as quite extraordinary and evidence of higher intelligence, since the branch is deliberately being used as a tool to change the behaviour of another animal, in this instance a Human (so that it goes away).

Orang-utans also use branches to swat away insects and fan themselves when they feel hot. They might pluck a stick and clean their teeth or ears and, when it pours torrential rain, they fabricate their own umbrella by balancing large leaves on their heads. If this does not work they may actually build a makeshift shelter. Daily, they also build a nest or platform on which to rest overnight. They sleep, like Humans, on their sides with their hands pillowing their head. Wallace observed that the Orang-utan, quite sensibly, 'does not leave his bed till the sun has well risen and has dried up the dew upon the leaves'.

Nest-building in itself may not seem amazing because lots of animals do it. But not many animals are as big as an Orang-utan. Indeed of all animals to live in trees, the Orang-utan is by far the largest. Adult males can weigh up to 97 kilograms and grow to a height of 1.4 metres, with an arm span of an extraordinary 2.4 metres; while the smaller females weigh about 40 kilograms and rarely grow above 1.2 metres. Possibly in response to their weight, females live and feed in the upper and middle canopies with the bigger males at the bottom, where their weight can be better supported by greater load-bearing branches. Building a nest to support such a weight is quite an engineering feat in anyone's language.

Orang-utans build new nests each day because they never seem to be in the same place. Since the 1970s, Orang-utans were considered to occupy stable foraging territories or harems, where the male's range overlapped that of one or two females. More recent observations dispute this and suggest that Orang-utans are semi-nomadic, just like the traditional peoples of tropical forests, following known routes of resource richness—perhaps from the fruiting trees of the hills, to the fresh shoots growing along the river banks. These routes are anchored by 'base camps'. Such camps could be a fruit-rich durian tree which has extraordinary pineapple-sized fruit, covered with conical spines and with a smell stronger than the smelliest cheese and a taste which definitely needs acquiring. Kept for some time, Wallace admitted that durian has 'a most disgusting odour' that is 'often so offensive that some persons cannot bear to taste it'.

Wallace himself was one of those people, until he sampled one straight from the forest floor. He was hooked.

'The pulp is the eatable part and its consistence and flavour are indescribable. A rich, butter-like custard highly flavoured with almonds gives the best general idea of it, but intermingled with it come wafts of flavour that call to mind cream cheese, onion sauce, brown sherry and other incongruities. Then there is a rich glutinous smoothness in the pulp which nothing else possesses, but which adds to its delicacy. It is neither acid, nor sweet, nor juicy, yet one feels the want of none of these qualitites, for it is perfect as it is. It produces no nausea or other bad effect and the more you eat of it the less you feel inclined to stop. In fact, to eat durians, is a new sensation worth a voyage to the East to experience.'

Irregularity is a characteristic of the tropical rainforest and it is one of the reasons frugivorous species, such as Orang-utans, need to forage over large terrains. The lack of regularity in flowering and sparse dispersal of each species of plant is compensated, in Borneo, by an enormous variety of species.

Orang-utans do not need to engage in territorial contests over resources, nor do they monopolise resources. Indeed, the presence of a large number of Orang-utans feeding peacefully together is not an atypical sight.[34] The ferocious-looking incisors of the Orang-utan are used, not for ripping away flesh from a fresh kill, not even for gouging at the flesh of a rival, but for gnawing away at the bark of certain trees. Orang-utans do eat meat, but only rarely and then usually insects.

Gisela Kaplan and Lesley Rogers noted wryly that a species without any noteworthy enemies, except for the now near-extinct Clouded Leopard and the python (and Humans), without deadlines to meet and buses to catch, without the urgency of a hunter or of prey being hunted, is usually not in a hurry.[34] Moreover, the Orang-utan, throughout its long history of existence, has been used to an abundance of food, unshared as it is in its relatively solitary existence in a large terrain teeming with the most diverse plant life and fruit on earth.

Compared with the oceanic island forests on the other side of the Makassar Strait, the continental Sunda Shelf forests are super-saturated in terms of species. Some of the tallest, densest, most luxuriant forests in the world are found here. In Borneo alone, there are some 3000 species of tree and 1200 species of orchid (not to mention other plant types). A single hectare of jungle may include 100 different tree types and the next hectare will add almost as many species to the list.

The predominance of the 250 or so different types of dipterocarp trees ('dipterocarp' means 'two-winged', named after their two-winged fruit) among the continuous canopy has led to the lowland vegetation being called dipterocarp forest. Dipterocarps do not disperse well over water and only six or so species have managed to disperse across the Makassar Strait to Sulawesi.

In southern Sarawak, Wallace found the super-abundance of plant life an inconvenience. After a few months at the coal mine, he decided to take a trip on a narrow tributary, to a Dayak house which was in close proximity to a mountain with an abundance of fruit and hence Orang-utans and birds. He travelled in a very small boat, carrying a cask of alcoholic arrack, only, of course, for the preservation of any animals he collected. On the banks Wallace spied all sorts of monkeys—macaques, leaf monkeys and the extraordinary long-nosed or Proboscis Monkey *Nasalis larvatus*, the outstanding feature of which is a huge, pendulous nose in males. Found only in Borneo, Prosboscis Monkeys are animals of the mangrove forests and feed on leaves and fruits. Their numbers, today, are also declining.

The stream became narrow and winding and Wallace could touch the vegetation on either side with his hands. Floating mats of grass blocked progress. Tangled branches and creepers built screens across the stream and screw-pines *Pandanus* sp., grew abundantly in the water, throwing a dense mop of blade-sharp leaves across the water.

After two days, he reached a landing place with a Dayak long-house, perhaps 80 metres long, nearby. It rose high above the ground on posts, with a wide veranda, containing baskets of human skulls and still wider platform of bamboo in front of it. Most of the Dayaks were away searching for edible birds' nest or beeswax and only a few old men and women with a lot of children remained. Wallace stayed here a few days and made forays out into the jungle searching for more Orang-utans.

One day he shot an Orang-utan which had been gorging, high up, on durians. As usual, Wallace shot the animal dead. He was more than a little excited, however, to discover that it appeared to be a different type, perhaps, he thought, a new species. It was a fully grown male but it did not have any sign of the lateral protuberances that uniquely grow on either side of the male Orang-utan's head.

Fully grown Orang-utan males develop extraordinary sized cheek pads and a full laryngeal sack, like an enormous double chin, which enables them to boom their calls out for miles. The development of these cheek pads is, however, not as much determined by age, as by context: the presence of other adult males will suppress or retard the development of the cheek pads. Wallace's new species was in all likelihood just a frustrated adult male.

Today the Orang-utan lives only in Borneo and Sumatra. The Sumatran Orang-utan is only slightly different from its Bornean cousin in that males do not develop such pronounced cheek pads and are smaller. But within Borneo there are differences between populations of Orang-utans. Most intriguing is the fact that the Orang-utans of south-western Borneo and Sumatra seem more closely related than the Orang-utans of Sabah and Sarawak—within Borneo. The fish fauna of the Kapuas River, the largest in Indonesia, in western Borneo also has fish almost identical to those found in the rivers of eastern Sumatra, but quite different from those of the Mahakam River in eastern Borneo, separated from the Kapuas River by a single mountain-top ridge. Eighteen thousand years ago these non-mountain climbing fish of western Borneo and eastern Sumatra shared the same catchment and swam in the same streams.

The similarity, generally, between the Orang-utans of Borneo and Sumatra is an indication of just how recently the area was a single, unbroken land mass, with the Wallace Line more or less forming the shoreline.

Wallace, however, had no knowledge of ice ages and lowered sea levels. As usual he resorted to the distribution of animals to unravel these geographical puzzles. Leaving out bats (which flew from place to place), Wallace found 48 species of mammals common to the Malay Peninsula and Sumatra, Borneo and Java. Among these were seven species of Primates, 'animals who pass their whole existence in forests, who never swim and who would be quite unable to traverse a single mile of sea; 19 Carnivora, some of which no doubt might cross by swimming, but we can not suppose so large a number to have passed in this way across a strait which, except at one point, is from 30 to 50 miles wide; and five hoofed animals, including the tapir, two species of rhinoceros and an elephant. Besides these there are 13 rodents and four Insectivora, including a shrew-mouse and six squirrels, whose unaided passage over 20 miles of sea is even more inconceivable than that of the larger animals'.

More specifically Wallace pointed out that 'Borneo is distant nearly 150 miles from Biliton [Belitung], which is about 50 miles from Banca and this 15 from Sumatra, yet there are no less than 36 species of mammals common to Borneo and Sumatra. Java again is more than 250 miles from Borneo, yet these two islands have 22 species in common, including monkeys, lemurs, wild oxen, squirrels and shrews. These facts seem to render it absolutely certain that there has been at some former period a connection between all these islands and the mainland and the fact that most of the animals common to two or more of them show little or no variation, but are often absolutely identical, indicates that the separation must have been recent in a geological sense; that is, not earlier than the Newer Pliocene epoch, at which time land animals began to assimilate closely with those now existing'.

Wallace focused on species of mammals which were shared across localities to make his comparison. Comparing species which are not shared is just as compelling. In contrast to old oceanic islands where virtually every mammal is unique, endemism on Sumatra is only five per cent (7 of 110 species), on Java, 12 per cent (7 of 61 species) and on Borneo, the largest of the Sunda Shelf islands, about 21 per cent (29 of the 124 indigenous species). The number of species on each island is also strongly correlated with island area: Borneo is bigger than Sumatra which is bigger than Java. On oceanic islands, in contrast, endemism is strongly correlated with age. Tellingly, most of the endemism on the Sunda Shelf islands is restricted to montane habitats, themselves 'islands'. Borneo, the most mountainous of the islands of the Sunda Shelf, for instance, has the highest level of endemism.

Even more startling is the fact that, in contrast to oceanic islands, where even small islands of 47 square kilometres, such as Ilin island in the Philippines, will have endemic species, small Sunda Shelf islands never have endemic species. They simply have reduced fauna drawn from the nearest larger island;[26] even sizeable islands such as Bangka and Belitung, which were part of the mainland during the last glacial period, have no endemic species. Indeed, islands which were connected to the mainland during the last glacial period need to be as large as Java before endemism extends even to a handful of species. Endemism on the Sunda Shelf islands rarely extends to the generic level.

Wallace found it remarkable that the Orang-utan, an animal

so large, so peculiar and of such a high type of form (in other words, like us), should be confined to two islands. Contemplating this on a Dayak veranda, next to a basket of human heads, he mused that 'we have every reason to believe that the Orang-utan, the chimpanzee and the Gorilla have also had their forerunners. With what interest must every naturalist look forward to the time when the caves and tertiary deposits of the tropics may be thoroughly examined and the past history and earliest appearance of the great man-like apes be at length made known'.

From the study of fossils, it has been determined that Primates evolved within the mammalian line about 63 million years ago in the Paleocene, when forms very similar to the present-day lemurs of Madagascar first appeared. These are known as the lower primates or prosimians. In the Oligocene, about 40 million years ago, the Primates split into two main lineages, the New World platyrrhine monkeys of South America, the living forms of which include marmosets, tamarins and howler monkeys; and the Old World catarrhine monkeys of Africa, today including the cercopithecoids (baboons and macaques) and the hominoids (apes and Humans).

Monkeys and apes may have evolved from nocturnal prosimians probably as a result of the evolution of colour vision that went with day-time activity.

The first member of the ape lineage, which includes Orang-utans and Humans, was *Proconsul*, appearing in the Miocene epoch 23–17 million years ago. *Proconsul* evolved in the eastern African rainforests of that time and from this beginning spread through Europe and Asia. From this line, the lesser apes—the gibbons and siamang—evolved, perhaps 17 million years ago. A fruit-eating hominoid, known as *Sivapithecus* evolved about 12 million years ago and is thought to be the ancestor of the Orang-utan. Back in Africa the same ancestor that gave rise to *Sivapithecus* evolved into a series of apes, the Gorilla and chimpanzee about eight and six million years ago respectively and then, perhaps as early as five million years ago, *Homo*. This ape evolved at a time when rainforests were dwindling and it was more of an advantage to be on two legs than four, roaming on the open plains. Meanwhile *Homo*'s close cousins the Orang-utans of Asia were well and truly ensconced in their ancient forest homes including the rainforests of Borneo, where the rainforests are the oldest remaining on Earth.

Orang-utans share 98 per cent of our genetic make-up. No other animal is as closely related to us except for chimpanzees, which share 99 per cent of our genetic make-up. Of course not all our genes are expressed, or expressed in the same way—otherwise *Homo*, Orang-utan and Chimpanzee would all look very much more alike. Orang-utans, however, share our diseases and live to about the same age Humans did before the advent of modern medicine. As we have seen, they also raise their young for long periods. They are also the only other Primate that has sex facing each other, with either the male or the female on top, or—unlike humans (except in the imagination)—swinging from the trees. Some researchers have suggested that Orang-utans may, in fact, be more similar to us than chimpanzees because they 'use' the same sub-set of genes.

The average density of Orang-utans varies to some extent according to habitat availability and altitude. With the exception of Mount Kinabalu, in Sabah, where the Orang-utan once existed, they are not found above 1500 metres. They are found in a variety of habitats such as swamp and secondary forests and in hilly forests in low numbers. The highest density of Orang-utans is reached in one locality in Sumatra where an abundance of fig trees supports five Orang-utans per two square kilometres.

Today, however, there are probably no more than 20 000 Orang-utans left. Their stronghold is in the tall lowland dipterocarp forests of Borneo. During Wallace's time dipterocarps were sprinkled throughout the lowlands of the Bornean forests but because their timber is prized, that situation is fast becoming a thing of the past. For Orang-utans, dipterocarps are a structural support, providing vertical highways to the various levels of the canopy.

The demise of the dipterocarp forests, being decimated at the rate of 100 000 square kilometres per year, means that in eastern Malaysia—Sabah and Sarawak—the Bornean rainforest may disappear almost entirely within the next five years and, with it, the Orang-utan.

CHAPTER 11

WALLACE IN WONDERLAND

I T WAS November 1861 and, after six years in the Malay Archipelago, Wallace was finally going home to London. Just one more stop remained, Sumatra, where Orang-utans had first been discovered. The mail-steamer from Batavia (Jakarta) dropped him off on the island of Bangka, adjacent to south-eastern Sumatra, where he took an open sailing boat to a small fishing village on the swampy coast and then a rowing boat through vast mangrove plains to Palembang 'a distance of nearly 100 miles by water' up the Musi River. The entire journey took him several days.

When he got to Palembang, he discovered an extraordinary city of houses projecting into the river on piles, or built upon great bamboo rafts so that the buildings rose and fell with the tide. The houses were moored by rattan cables to the shore or to piles. 'The natives are true Malays,' Wallace observed with his typical dry sense of humour, 'never building a house on dry land if they can find water to set it in and never going anywhere on foot if they can reach the place by boat.'

Wallace's enquiries about the Orang-utan only brought him blank looks. To his surprise no-one had heard of such an animal and he was forced to conclude that it did not inhabit the great forest-plains in the east of Sumatra where he most expected to find it. In many respects Wallace solved this apparent puzzle by observing that the elephant, 'the other great Mammalia of Sumatra' was also much more scarce than it had been a few years before, 'and seems to retire rapidly before the spread of cultivation'. In Sumatra, Orang-utans are still confined to the very north of the island in Gunung Leuser Park, an area that Wallace did not visit.

His brief Sumatran visit was not, however, in vain. One morning while waiting for a boat to be made water-tight to take him back to the coast, he sent a couple of his hunters out to shoot the birds of the area surrounding Palembang. While he was eating breakfast they returned with a large male specimen of the Great Pied Hornbill *Buceros bicornis*. This was a turkey-sized black and cream bird with a large, striking, decurved bill, bearing a characteristic 'casque', a structure that projects from the top of the bill (*buceros* is derived from the latin *bucerus* meaning 'having ox's horns').

One of Wallace's hunters assured him that they had shot the male hornbill while feeding the female, which the male had imprisoned by plastering her up in a tree hole. Wallace's curiosity was fired and he raced to the place with a host of villagers following.

'After crossing a stream and a bog, we found a large tree leaning over some water and on its lower side, at a height of about 20 feet, appeared a small hole and what looked like a quantity of mud, which I was assured had been used in stopping up the large hole. After a while we heard the harsh cry of a bird inside and could see the white extremity of its beak put out. I offered a rupee to any one who would go up and get out the bird, with the egg or young one, but they all declared it was too difficult and they were afraid to try. I therefore very reluctantly came away. In about half an hour afterward, much to my surprise, a tremendous loud hoarse screaming was heard and the bird was brought to me, together with a young one which had been found in the hole. This was a most curious object, as large as a pigeon, but without a particle of plumage on any part of it. It was exceedingly plump and soft and with a semi-transparent skin, so that it looked more like a bag of jelly, with head and feet stuck on, than like a real bird. The extraordinary habit of the male in plastering up the female with her egg and feeding her during the whole time of incubation and till the young one is fledged, is common to several of the large hornbills and is one of those strange facts in natural history which are "stranger than fiction".'

Stranger still is the fact that it is the female which locks herself away and seals the entrance of the tree hole leaving a narrow vertical slit, through which she and later the chicks, receive food from the male and out of which food remains and droppings are voided.[35]

Choosing an appropriate hole, however, is done by both male and female hornbill with a thoroughness that is only matched by human home-buyers. The pair, particularly the female, go around poking their heads into each nook and cranny and pecking around within them. They check that the site is close to food, defendable against other hornbills and other hole-living animals and has a comfortable microhabitat in which to incubate the eggs and rear a young family. The entrance hole must be small enough to be sealed up, but large enough to admit the female and must have a rim on which sealing material can be easily applied. An even floor, a high ceiling, thick and robust walls, lack of cluttering protrusions and good ventilation are all good selling points.

When the female finally begins to spend more and more time inspecting a specific hole, the male knows it is time to nest. He begins to bring her lining and sealing materials with the food he has brought since courting and home inspection began. Then copulation takes place, consummating the relationship. The female continues to seal the entrance from the outside and inside, applying mud or sticky foodstuffs and even her own faeces once she has locked herself inside. Sealing material is held in the bill tip and squeezed out the sides and then applied using the broad, flattened sides of her bill like a spatula. The male also brings material to line the nest and to level out the floor.

Finally sealed inside, the female does not lay her eggs immediately. Instead she monitors the security of the nest and the diligence of her mate in providing for her. She will break out of the nest if anything is not up to standard and safe and secure. She has good reason to be so fussy because once egg laying begins, she puts her life literally in the wings of her mate, shedding all her flight feathers. Totally vulnerable, she now has more energy to devote to the task of rearing her young and, without the long feathers, more room to manoeuvre in the cramped tree hole. The feathers grow back while she incubates the eggs and broods the chicks.

Incubation may last up to 42 days[35] and the female spends long periods with her eye pressed to the entrance slit. If a predator comes to the entrance she bolts up to the top of the hole where she secretes herself against the high ceiling. In this way large chicks and the female can seemingly disappear from sight.

Sanitation in such a cosy home is dealt with in a most hygienic (if absurd) fashion. The female turns away from the entrance, reverses up to the narrow opening, carefully positions her anus and then defecates with considerable force sending out a spray of material. Soiled material from young chicks is also tossed out by the mother.

Wallace's observation that the chick looked like a big bag of jelly was not far wrong. After hatching, hornbill chicks develop an extraordinary air sac under the skin which spreads over most of the body. Exactly why the chicks develop into living balloons is not yet clear.

The hornbill home becomes increasingly noisy as the chicks develop. Their strident cries for food are interspersed with a loud, crescendo acceptance call on taking food from the male. Like their mother, they often participate in a monocular scan of the outside world. Perhaps not surprisingly, the female will leave the nest at around this stage, flying off rather stiffly after breaking down the nest seal with persistent hard pecking, with the assistance of the young. Only as much as is absolutely necessary is removed since the nestlings need to reseal the nest entrance until they are fully developed.

Fifty-four species of hornbill with the familiar African and Asian distribution are currently recognised. At least eight species are found in the Sunda Shelf forests, seven endemic species exist in the Philippines, two endemic species in Sulawesi and one species, the Papuan Wreathed Hornbill *Aceros plicatus* even extends to the Solomons.

The Great Pied Hornbill and its spongy youngster was perhaps Wallace's last major natural history observation before quitting the Malay Archipelago forever. It must have been a poignant time. A time also tinged with a little irony since this, his last trip in the archipelago, was once again in a ricketty boat, 'water-tight' in name only, but nevertheless his lifeline to the relative safety of the mail-steamer, 'that highest triumph of human ingenuity, with its little floating epitome of European civilization'. Wallace had had a great deal to do with rafts and boats in this nation, more water than land.

On Timor, for instance, in May 1859—18 months earlier—he had taken a deep coffin-like boat from Semao, a small offshore island, to bring him back to Kupang. The boat was filled with his baggage and with vegetables and fruit for the Kupang market.

'When we had got some way across into a rather rough sea, we found that a quantity of water was coming in which we had no means of bailing out. This caused us to sink deeper in the water and then we shipped seas over our sides and the rowers, who had before declared it was nothing, now became alarmed and turned the boat round to get back to the coast of Semao, which was not far off. By clearing away some of the baggage, a little of the water could be bailed out, but hardly so fast as it came in and when we neared the coast we found nothing but vertical walls of rock, against which the sea was violently beating. We coasted along some distance till we found a little cove, into which we ran the boat, hauled it on shore and, emptying it, found a large hole in the bottom, which had been temporarily stopped up with a plug of cocoa-nut, which had come out. Had we been a quarter of a mile further off before we discovered the leak, we should certainly have been obliged to throw most of our baggage overboard and might easily have lost our lives. After we had put all straight and secure we again started and when we were half-way across, got into such a strong current and high cross-sea that we were very nearly being swamped a second time, which made me vow never to trust myself again in such small and miserable vessels.'

Famous last words! Less than a year later, in April 1860, Wallace was enthusiastically building his own perahu on

WALLACE IN WONDERLAND

Gorong, one of a string of pirate-inhabited islands to the east of Seram, Maluku. In it he would take a nightmare journey lasting months to the New Guinean offshore island of Waigeo. Wallace's crew, who barely lived through the trials of the journey, blamed the misfortunes on the fact that an important ceremony had not been carried out on the little vessel.

They were referring to the fact that, as all locals knew, objects which have character, power and great use, such as a boat, have an intrinsic spirit. The effort of a craftsman in fashioning a boat transfers power into it, so that it becomes a force to be reckoned with.[29] The danger of ocean voyaging can be partly neutralised by magic that is built into the canoe during its construction. A canoe built without magic is a dangerous thing.

Wallace had bought the shell of a perahu, which no doubt had been imbued with its own life-force through a ceremony which involved the implantation of gold—the magical symbol of semen—wrapped in a piece of white cloth, in a hole in the deepest, firmest place in the most important joint: the place where the (male) stempost is married to the (female) hull.[29]

Wallace intended to pay several expert boat-builders from the nearby Kai islands to fit out the inside of this boat. However he found that, unless he was constantly on the spot, very little work was done. He was obliged to give up collecting to do the inside work himself, much to the amazement of the locals who were surprised to see a white man at work. The ever-resourceful Wallace had brought 'a few' tools of his own, including a small saw and some chisels which were 'of the best London make'.

Wallace noticed that the locals were 'much astonished' at the novel arrangements he was making inside the perahu. The real reason for their astonishment was more likely Wallace's dangerous disregard of the power of the perahu's conservative spirit which would spitefully seek reckoning for any advantage such innovations might make.

Wallace launched and loaded his boat and then 'actually set

sail the next day (May 27th) much to the astonishment of the Goram people, to whom such punctuality was a novelty'. Again the real reason for their astonishment was more likely the fact that they had never seen a person so goad a spirit by adding insult to the injury of his innovations: Wallace's perahu was wrecklessly launched without the proper launching ceremony that would keep it safe while at sea. Just before first launching the perahu, a hole should have been gouged at an angle through the keel. The chips from this, the boat's 'navel', should have been kept in a bottle of coconut oil and hung from a beam in the finished hull, much like the custom of keeping a dried piece of umbilical cord of a child to ward away sickness or danger.

Wallace's crew consisted of three men and a boy plus his own two 'lads' from Ambon. On the way to Seram, he stopped at the black-marketeering island of Kilwaru where New Guinean slaves, opium, pearls, birds of paradise and anything or anybody from China to New Guinea could be bought and sold. Wallace bought a variety of goods for barter and two muskets against pirate attacks. After a rocky voyage he reached the eastern tip of Seram in heavy seas which caused the perahu to roll about. The crew scrambled to get the perahu inside some reefs and safely anchored overnight. Throughout the night the boat rolled and jerked uneasily.

The next morning Wallace discovered his crew had abandoned ship. 'As I had treated my men with the greatest kindness and had given them almost everything they had asked for, I can impute their running away only to their being totally unaccustomed to the restraint of a European master and for some undefined dread of my ultimate intentions regarding them.' No doubt there was 'undefined dread' involved, but it probably had little to do with Wallace and a great deal to do with the luckless, perhaps even malevolent, boat.

It took Wallace several days to gather together a few men to take him to the next town along Seram's northern coast, Wahai. In the meantime, stranded near a small, isolated village, Wallace had the opportunity to investigate the process of sago making. Today, as then, sago flour is a staple of all the lowland peoples in the region. The swamp-loving Sago tree *Metroxylon sagu* is a palm a little like a Coconut tree but with spiny leaves which completely cover the trunk. It takes the tree 15 years to build up the energy required to flower, after which it dies.

When sago is to be made, a full-grown tree is selected just before it goes to flower. The tree is felled close to the ground, the leaves stripped off and the bark peeled away revealing the nutrient-rich pith which has the same general consistency and colour as dried apple, but with woody fibres running through it. Wallace observed how the villagers chopped the pith free of the woody fibres with a specific sago club. The pith was then carried to the nearest fresh water and kneaded, strained and soaked to wash out the starch. Boiled with water this starch makes a thick, glutinous mass which, Wallace observed, was eaten with salt, limes and chilli. Sago bread, cooked in a clay oven over clean coals was also made. Wallace ate these with butter and observed that 'when made with the addition of a little sugar and grated cocoa-nut are quite a delicacy'. Dried sago lasts for years.

Wallace calculated that 'a good-sized tree will produce 30 tomans or bundles of 30 pounds [13.6 kilograms] each and each toman will make 60 cakes of three to the pound. Two of these cakes are as much as a man can eat at one meal and five are considered a full day's allowance; so that reckoning a tree to produce 1800 cakes, weighing 600 pounds, it will supply a man with food for a whole year. The labour to produce this is very moderate. Two men will finish a tree in five days and two women will bake the whole into cakes in five days more; but the raw sago will keep very well and can be baked as wanted, so that we may estimate that in ten days a man may produce food for the whole year'. Wallace had particular reason to be respectful of the Sago palm since all the insect boxes he filled were made from the midribs of the palm's enormous leaves.

Perhaps it was Wallace's incorrigible curiosity that, after a few days, made the villagers finally provide him with a crew to take him away to Wahai. The main Dutch settlement on Seram, it was almost due south of the western tip of New Guinea and its offshore islands of Misool and Waigeo, Wallace's destination. In Wahai, Wallace obtained four men to crew for him in the spirited vessel. Wallace summarised the eventful but disastrous next three months.

'Looking at my whole voyage in this vessel from the time when I left Goram [Gorong] in May, it will appear that my experiences of travel in a native prau have not been encouraging. My first crew ran away; two men were lost for a month on a desert island; we were ten times aground on coral reefs; we lost four anchors; the sails were devoured by rats; the small boat was lost astern; we were 38 days on the voyage home, which should not have taken 12; we were many times short of food and water; we had no compass-lamp, owing to there not being a drop of oil in Waigiou [Waigeo] when we left; and to crown all, during the whole of our voyages from Goram by Ceram to Waigiou and from Waigiou to Ternate, occupying in all 78 days, or only 12 days short of three months (all of which was supposed to be the favourable season), we had not one single day of fair wind! We were always close braced up, always struggling against wind, tide and leeway and in a vessel that would scarcely sail nearer than eight points from the wind. Every seaman will admit that my first voyage in my own boat was a most unlucky one.'

It was this voyage, ending in August 1860, that finally turned Wallace's thoughts to home: just as, ironically, it was a voyage in a native perahu from Makassar to Aru in 1856 that kept him in the Malay Archipelago for a further five years. If there was a particular high spot in all his travelling, the voyage to the Aru islands on the Sahul Shelf of Australia was it.

'I brought away with me more than 9000 specimens of natural objects, of about 1600 distinct species. I had made the acquaintance of a strange and little-known race of men; I had become familiar with the traders of the Far East; I had revelled in the delights of exploring a new fauna and flora, one of the most remarkable and most beautiful and least known in the world; and I had succeeded in the main object for which I had undertaken the journey—namely, to obtain fine specimens of the magnificent birds of paradise and to be enabled to observe them in their native forests. By this success I was stimulated to continue my researches in the Moluccas and New Guinea for nearly five years longer and it is still the portion of my travels to which I look back with the most complete satisfaction.'

Surely it was this journey, ending in July 1857—which came directly after he had taken the fortuitous detour via Bali and Lombok, across what was to become the Wallace Line—that inspired Wallace to write to Darwin, in early 1858, with a succinct explanation of evolution by natural selection.

In Wallace's time, the voyage to the Aru islands was looked on as a wild and romantic expedition, full of novel sights and strange adventures. By today's standards things are not so much different. From Ambon, a freighter might leave for the Aru islands once every month. Even from the nearby Kai islands, where a motor launch can do the trip in about 12 hours, a wait of two or three weeks is not uncommon.

Wallace did the journey as a passenger in a large Bugis perahu, a vessel shaped something like a Chinese junk. These vessels, crewed by about 50 men, only made the journey from Makassar, Sulawesi, once a year, owing to the monsoons—the boats leaving in December or January at the beginning of the western monsoon and returning in July or August at the full strength of the eastern monsoon. Despite the infrequency of the journey, the Aru islands were very valuable. Shiploads of tripang, or sea-slug ('looking like sausages which have been rolled in mud and then thrown up the chimney') and edible bird's nests were brought back for the gastronomic enjoyment of the Chinese; and pearls, mother-of-pearl, tortoise-shell and birds of paradise found their way to Europe.

On the perahu, Wallace occupied a little thatch compartment about 'six-and-a-half feet long by five-and-a-half wide' on the deck. He said of it that 'it was the snuggest and most comfortable little place I ever enjoyed at sea'. Built only with natural vegetable fibres, with no paint, tar, oil or varnish (which made him 'qualmish'), the vessel's smell made Wallace 'recall quiet scenes in the green and shady forest'.

It was a fairytale voyage with the wind occasionally blowing from the right direction and carrying the lumbering vessel at five knots per hour—as fast as it could go. In the evenings Wallace gazed into the rushing water where eddying streams of phosphoric light gemmed and whirled. It resembled more than anything else 'one of the large, irregular, nebulous star-clusters seen through a good telescope, with the additional attraction of ever-changing form and dancing motion'.

He noted with typical humour that the crew was numerous enough for a vessel of 700 tons instead of the 70 tons of this vessel. 'They have it very much their own way and there seems to be seldom more than a dozen at work at a time. When any thing important is to be done, however, all start up willingly enough; but then all think themselves at liberty to give their opinion and half a dozen voices are heard giving orders and there is such a shrieking and confusion that it seems wonderful any thing gets done at all.'

Despite this there were no quarrels and everyone passed the time amicably, chewing betel nut or talking or sewing or sleeping. The time itself was kept in a most ingenious manner. In a half-filled bucket of water was placed a well-scraped half of a coconut shell which had a tiny hole at its base. The shell floated on the water, slowly letting in a small trickle of water. This gradually filled the shell so that exactly at the end of an hour, plump it went to the bottom. Wallace, of course, measured it and found that the water clock varied less than a minute from one hour to the next.

Christmas Day 1856 came and went, Wallace celebrating it with an extra glass of wine and a fine view of Buru. The vessel continued, sailing past the perfect cone of the Banda volcano and anchoring for a short stay among the Kai islands with their dazzling white beaches and water as transparent as crystal. Wallace was enthralled. 'The scene was to me inexpressibly delightful. I was in a new world and could dream of the wonderful productions hid in those rock forests and in those azure abysses. But few Europeans' feet had ever trodden the shores I gazed upon; its plants and animals and men were alike almost unknown.' Wallace well knew that he was entering the realm of New Guinea and Australia.

The Kai islanders were famous across the archipelago for their pre-eminent skills in boat-building. The fact was not lost on Wallace who noticed that the small, beautiful canoes were built 'without a nail or particle of iron being used and with no other tools than axe, adze and auger'. So well-honed was the skill that 'the best European shipwright can not produce sounder or closer-fitting joints'. That the evolution of such a skill had taken millennia is self evident. It is, however, the rigging that gives a real clue to the antiquity of the art form. The rigging of the traditional vessels of Indonesia is fundamentally different from modern craft. It is designed to steer a vessel without a rudder, a little like a windsurfer. In fact the basic principles seem to have been worked out, not on canoes, but on rafts.[29] And rafts, it seems, have an awesome antiquity.

In 1991, a joint Indonesian and Dutch palaeontological

expedition discovered human artefacts in a layer fairly accurately dated at around 730 000 years before present.[65] What is surprising about this finding is that it is on the oceanic island of Flores. Even during low sea levels the human community (probably *Homo erectus*) who made the artefacts, would have had to have rafted there, making them the earliest known mariners.

Thirty hours from the Kai islands, Wallace's vessel anchored in the harbour of Dobbo at the northern end of the flat, forested Aru islands. Wallace was delighted with the trip and rated 'the semi-barbarous prau as surpassing those of the most magnificent screw-steamer, that highest result of our civilization'.

Even the rather miserable thatch shed Wallace found to put his things in could not dampen his spirits. First thing the next morning he was off to explore the virgin forests of Aru. Less than a kilometre out of Dobbo, Wallace was pushing through trackless forest that even his servants had trouble chopping a path through. At the end of the day he had gone less than four kilometres but had, nevertheless, managed to capture about 30 species of butterflies 'more than I had ever captured in a day since leaving the prolific banks of the Amazon and among them were many most rare and beautiful insects, hitherto only known by a few specimens from New Guinea'.

Actually there were many places in the Malay Archipelago that Wallace described in such a way. In Singapore, at the beginning of his journey, he obtained no less than 700 species of beetles collected in one patch of jungle 'not more than a square mile in extent'. Wallace had rarely, if ever, met with so productive a spot. That was until he went to Borneo where he found about 24 new species of beetles a day and on one day, 76 different kinds. Altogether in Borneo he collected about 2000 different kinds of beetles 'on scarcely more than a square mile of ground'. In Banda, too, he collected more large and brilliant species than he had ever done before in a short time. In Bacan, Wallace had never seen beetle so abundant, obtaining 70 distinct species of beetle in 'a place where a new clearing was being made in the virgin forest ... It was a glorious spot and one which will always live in my memory as exhibiting the insect-life of the tropics in unexampled luxuriance'.

He found much to delight him both in the wildlife and the people. As usual when he was arranging his insects in their boxes, he was surrounded by a crowd of wondering spectators. One day, in Bacan, he showed one of them how to look at a small insect with a hand-lens 'which caused such evident wonder that all the rest wanted to see it too. I therefore fixed the glass firmly to a piece of soft wood at the proper focus and put under it a little spiny beetle of the genus *Hispa* and then passed it round for examination. The excitement was immense. Some declared it was a yard long; others were frightened and instantly dropped it and all were as much astonished and made as much shouting and gesticulation, as children at a pantomime, or at a Christmas exhibition of the oxyhydrogen microscope. And all this excitement was produced by a little pocket-lens an inch-and-a-half focus and therefore magnifying only four or five times, but which to their unaccustomed eyes appeared to enlarge a hundredfold'.

Since Wallace's time, the most comprehensive attempt at discovering the insect abundance and diversity in tropical rainforests was carried out in northern Sulawesi in only 1985[38] during a project appropriately code-named Project Wallace. In contrast to Wallace's steady, 'hands-on' approach, Project Wallace employed a battery of modern-day methods on a 500 hectare plot (about five square kilometres), including light traps where insects are lured by bright light, fogging with insecticide, pitfall traps in the ground and flight intercept traps in the canopy. Over a period of a year, several million insects were caught including 1 172 000 beetles of around 6000 species (and still counting).

An extra-terrestrial ecologist trying to summarise the species on planet Earth would surely have to say that virtually all life on Earth was insect. But as much as Wallace loved insects it was, finally, a bird that made his entire trip worthwhile.

Wallace was lost in admiration. In his hands lay a legendary bird, a bird that no European had ever seen in the wild, even though mutilated specimens had been described by Linnaeus a century before. 'One of the most perfectly lovely of the many lovely productions of nature', Wallace held the freshly killed body of the King Bird of Paradise *Cicinnurus regius*. Only a young child with a full palette of primary colours and a clean sheet of paper could possibly have conjured such a colour scheme. And only Wallace could provide the vivid, jubilant prose to describe it.

'It was a small bird, a little less than a thrush. The greater part of its plumage was of an intense cinnabar red, with a gloss as of spun glass. On the head the feathers became short and velvety and shaded into rich orange. Beneath, from the breast downward, was pure white, with the softness and gloss of silk and across the breast a band of deep metallic green separated this colour from the red of the throat. Above each eye was a round spot of the same metallic green; the bill was yellow and the feet and legs were of a fine cobalt blue, strikingly contrasting with all the other parts of the body. Merely in arrangement of colours and texture of plumage this little bird was a gem of the first order, yet these comprised only half its strange beauty. Springing from each side of the breast and ordinarily lying concealed under the wings, were little tufts of grayish feathers about two inches long and each terminated by a broad band of intense emerald green. These plumes can be raised at the will of the bird and spread out into a pair of elegant fans when the wings are elevated. But this is not the only ornament. The two middle feathers of the tail are in the form of slender wires about five inches long and which diverge in a beautiful double curve. About half an inch of the end of this wire is webbed on the outer side only and coloured of a fine metallic green and being curled spirally inward, form a pair of elegant glittering buttons, hanging five inches below the body and the same distance apart. These two ornaments, the breast-fans and the spiral-tipped tail-wires, are altogether unique, not occurring on any other species of the eight thousand different birds that are known to exist upon the earth.'

Today 9672 species of birds are recognised worldwide.[23] Wallace knew of 17 birds of paradise. Today 43 are recognised.

When the first Europeans reached the Spice Islands in the early sixteenth century in search of cloves and nutmeg, they were presented with the dried skins of birds so gorgeous as to be revered. The Portuguese, finding that they had no feet or wings, named them Birds of the Sun, while the Dutch called them Paradise Birds. Because no-one had apparently ever seen the birds alive it was said that they lived in the air, always turning to the sun and never lighting on the earth till they die. Indeed, in 1760, Linnaeus named the large Greater Bird of Paradise *Paradisaea apoda*, the Footless Paradise Bird, a name the bird keeps to this day.

It was the way that the birds were preserved that gave them this ethereal aspect. Wallace described it matter-of-factly. 'The native mode of preserving them is to cut off the wings and feet and then skin the body up to the beak, taking out the skull. A stout stick is then run up through the specimen coming out at the mouth. Round this some leaves are stuffed and the whole is wrapped up in a palm spathe and dried in the smoky hut. By this plan the head, which is really large, is shrunk up almost to nothing, the body is much reduced and shortened and the greatest prominence is given to the flowing plumage.'

One hundred years after Linnaeus described the Greater and the King Bird of Paradise, Wallace became the first European to see the legendary birds in the wild.

As much as Wallace was awed by being the first European to see living legends, the native Aru people were awed by the sight of their first white man. Wallace stayed in a house together with four or five families and from six to a dozen visitors. 'They kept up a continual row from morning to night—talking, laughing, shouting without intermission—not very pleasant but interesting as a study of national character.' It was not long before Wallace discovered that he was the focus of the row and the steady stream of visitors. It was the first time a real white man had come to stay with them. Wallace found this very flattering. He recalled how, in London, he himself had been 'one of the gazers at the Zulus and the Aztecs'. Now the tables were turned 'for I was to these people a new and strange variety of man and had the honour of affording to them, in my own person, an attractive exhibition gratis ... It was only while gazing at me that their tongues were moderately quiet, because their eyes were fully occupied'. As Wallace's stay lengthened his hosts became more and more curious about his activities. Why did he so carefully preserve all the insects, birds and other animals and put them in boxes? Wallace's reply, that they would be stuffed and made to look alive so that people would go to look at them, was not to be believed.

Wallace not only sought to gain specimens of the birds of paradise but also to gain some knowledge of their habits. He discovered that the Greater Bird of Paradise, at least, seasonally participated in extraordinary 'dancing-parties' within a specific communal display area, known as a 'lek'. Typically leks are places where the branches of a tree are gently sloping, providing plenty of room for display and covered with a thick umbrella of canopy. The leks, which are used year after year, are occupied by around 'a dozen or twenty' adult males.

The Greater Bird of Paradise, Wallace said, 'is nearly as large as a crow and is of a rich coffee-brown colour. The head and neck is of a pure straw yellow above and rich metallic green beneath. The long plumy tufts of golden-orange feathers spring from the sides beneath each wing and when the bird is in repose are partly concealed by them. At the time of its excitement, however, the wings are raised vertically over the back, the head is bent down and stretched out and the long plumes are raised up and expanded till they form two magnificent golden fans, striped with deep red at the base and fading off into the pale brown tint of the finely divided and softly waving points. The whole bird is then overshadowed by them, the crouching body, yellow head and emerald green throat forming but the foundation and setting to the golden glory which waves above. When seen in this attitude, the bird of paradise really deserves its name and must be ranked as one of the most beautiful and most wonderful of living things.'

The Greater Bird of Paradise, along with virtually all other birds and all mammals on Aru, is shared with New Guinea, a fact which made Wallace realise that Aru had been recently connected with New Guinea (just as the fauna of the Greater Sunda Islands had made him realise they had been recently connected to the Asian mainland). Wallace knew of only two mammals present in Aru: a 'true kangaroo' and a bandicoot. Fourteen species of indigenous mammals have now been recorded from Aru, including carnivorous marsupials, kangaroos, bandicoots, cuscus, possums and native rodents—all the same species as occur on the mainland. The difference between this 7700 square kilometre, offshore island and oceanic Seram, more than twice the size but the same distance from mainland New Guinea as Aru, could not be more striking, mirroring the differences between Sulawesi and the offshore islands of Borneo, Sumatra and Java. Seram has only five indigenous mammals and of these the Northern Common Cuscus and the Common Spotted Cuscus are likely to have been introduced prehistorically,[19] leaving one bandicoot and two native rodents—all endemic.

In Singapore on his way home, Wallace found two caged

Lesser Birds of Paradise *Paradisaea minor*, smaller than the Greater Bird of Paradise but similar looking. Since they seemed healthy and fed voraciously on cockroaches, rice and bananas he bought them for the extraordinary price of 100 pounds, the equivalent, today, of several thousands of dollars. Ever the eccentric, Wallace stopped at Bombay for a week to lay in supplies of bananas for his birds and at Malta for two weeks, spending the time filling several biscuit tins full of live cockroaches (complaining that on modern-day steamers cockroaches were becoming scarce!). The birds arrived in London in perfect health and lived for two years in the Zoological Gardens.

EPILOGUE

They have all lately gone.

Alfred Russel Wallace. The Malay Archipelago, *1869.*

I<small>N</small> 1992, I had the opportunity to take an expedition on a wooden boat—a tall brigantine—into remote parts of eastern Indonesia. The ship sailed from Ambon, moving gracefully along its picturesque harbour. In 1857, Wallace described the harbour as being like a fine river, the clarity of the water affording him one of the most astonishing and beautiful sights he had ever beheld. 'The bottom was absolutely hidden by a continuous series of corals, sponges, actiniae and other marine productions, of magnificent dimensions, varied forms and brilliant colours. The depth varied from about 20 to 50 feet and the bottom was very uneven, rocks and chasms and little hills and valleys, offering a variety of stations for the growth of these animal forests. In and out among them moved numbers of blue and red and yellow fishes, spotted and banded and striped in the most striking manner, while great orange or rosy transparent Medusae floated along near the surface. It was a sight to gaze at for hours and no description can do justice to its surpassing beauty and interest. For once, the reality exceeded the most glowing accounts I had ever read of the wonders of a coral sea. There is perhaps no spot in the world richer in marine productions, corals, shells and fishes than the harbour of Amboyna.'

I looked down eagerly as our ship sailed out of the harbour, but through the cloudy water saw only a rubbish tip of lifeless coral. The rest had been dredged and used as fill for Ambon's construction boom.

Along the north coast of Seram, the wind, which had been as contrary as Wallace found it, finally filled the sails and the vessel sliced through the waves, slapping the water. Heavily forested and mountainous, majestic Seram is still perhaps the only place left in South-East Asia where you can walk continuously through undisturbed lowland forest, through shrubby *Rhodendron* and into high altitude tree-fern grassland, even though logging now continues apace.

Once the vessel cleared Seram we headed for Sorong on the tip of Irian Jaya, passing Misool to our east. We would stop here on our return journey. Wallace on his luckless voyage to Waigeo never made it to Misool despite being close enough to swim to it. The wind and current, instead, whipped his vessel toward shipwreck on some nearby coral.

I volunteered to stand watch at the bow. Even though the vessel had modern radar and sonar, large logs and small vessels could still slip through. Bow-watch from 4 a.m. to 8 a.m. was my favourite. The bio-luminescence of millions of living organisms was as magical now as when Wallace observed it, swirling crazily in a watery cosmos. Like staring into an open fire it was hypnotic, until a pearly dawn put out the lights.

From Sorong we returned toward Misool, slipping through the narrow Sele Straits between the mainland and the island of Salawati. We sailed all night passing by the roaring flames, like erupting volcanoes, of Salawati's oil wells.

Next morning we reached Misool again. Large jagged hills of limestone stuck out of the sea like teeth. Manoeuvring carefully, we anchored the brigantine in a glorious little bay, near a small village. An anachronism in the 1990s, it was probably the first time in 200 years, even 300 years, that a Tall Ship had visited these waters. The entire village turned out to greet us. It was the first time that many of them had seen white faces and the din as some two hundred people talked, laughed and shouted at once was deafening.

A dug-out canoe with an oily outboard motor took us to the site of an inland village. All that remained of it, however, was a few pieces of pottery and a swathe of fire-induced coarse grass. I was reminded of Wallace who described such areas as 'absolutely destitute of interest for the zoologist'.

On the nearby Kai islands, which lives up to its reputation as having some of the most glistening white beaches in the world, my heart sank as I saw that the coarse grass now dominates the hills that Wallace described as inexpressibly beautiful and 'clothed throughout with a most varied and luxuriant vegetation'. No-one will ever know what biological treasures existed here.

The ship sailed south-west through a deep blue, almost black, sea. We were over the abyssal depths where the Australian tectonic plate plunges under the Asian one. On a whim, we took a swim in this, one of the deepest oceans in the world, floating six kilometres above the sea floor, the ship rocking on the swell, without anchor. Even here the presence of Australia was felt in the quality of the atmosphere, hazy with distant smoke. Ancient people, more attuned than I to their environment, must always have known that she was there. As we neared the continental shelf, the sea turned aquamarine, sprinkled with the red dust of my arid land.

My trip ended in Darwin but it is worthwhile briefly revisiting parts of Wallace's. In Sulawesi he spent days plodding the rocky rainforests north of Makassar (Ujung Pandang), with his net and collecting bottles, plucking up a beautiful little green-and-gold-speckled weevil here, a pale blue and black butterfly there. Here some of the finest butterflies in the world, the bird-winged butterflies with wing spans up to 20 centimetres, wheeled through the thickets with a strong sailing flight.

Wallace frequently visited a series of waterfalls along a rocky creek and when the sun shone hottest about noon, butterflies flocked to the dappled, sandy bank of the pool below the upper fall. It was a fairytale sight with parties of gay butterflies—orange, yellow, white, blue and green—which, when disturbed, rose into the air by hundreds forming clouds of variegated colours. Here, also, Wallace obtained an insect which he had hoped but hardly expected to meet with—the magnificent *Papilio androcles*, one of the largest and rarest swallowtailed butterflies. This beautiful creature has long white tails which flicker like streamers in flight. When it settles it carries them delicately raised upwards as if to preserve them from injury, like a woman raising long, white, silk skirts from a muddy floor. The butterfly, known today as *Graphium androcles*, has not been seen for decades.[79]

On the tip of the northern peninsula of Sulawesi, Wallace visited 'a place celebrated for Maleo, as well as for Babirusa and Sapi-utan'. He watched, with evident delight, as the proud Maleo strutted down to the hot sands of the beach. Today the area is a tiny 87 square kilometre national park, Tangkoko Reserve. The Babirusa has gone and the dwindling populations of anoa and Maleo cannot survive in such a scrap of forest.

In Sarawak where Wallace 'collected' his Orang-utans, no Orang-utans exist anymore. The forested sites where he found beetles in their infinite variety, in Singapore, in Sarawak and in Ambon, no longer exist.

This book has been a great joy and a great sadness to write. If Wallace were to retrace his journey today, in truth he would have found little to inspire him. And without inspiration it is doubtful he would have developed the theory of evolution and the intellectual mystery of The Wallace Line. We lose more than wildlife in destroying nature, we lose our humanity.

REFERENCES

1 Agassiz, L. (1857) Contributions to the Natural History of the United States of America, Vol.1 In E. Mayr (1976) Evolution and the Diversity of Life: selected essays. Belknap Press, Cambridge.
2 Audley-Charles, M.G. (1981) Geological History of the Region of Wallace's Line In T.C. Whitmore (ed.) (1981) Wallace's Line and Plate Tectonics. Clarendon Press, Oxford.
3 ——(1988) Evolution of the Southern Margin of Tethys (North Australian Region) from Early Permian to Late Cretaceous In M.G. Audley-Charles & A. Hallam (eds) (1988) Gondwana and Tethys. Geological Society Special Publication No. 37. Oxford University Press, Oxford.
4 ——Ballantine, P.D. & Hall, R. (1988) Mesozoic-Cenozoic rift-drift sequence of Asian fragments from Gondwanaland. Tectonophysics 155, 317:330.
5 Audley-Charles, M.G., Hurley, A.M. & Smith, A.G. (1981) Continental Movements in the Mesozoic and Cenozoic In T.C. Whitmore (ed.) (1981) Wallace's Line and Plate Tectonics. Clarendon Press, Oxford.
6 Bates, H.W. (1864) The Naturalist in the River Amazon In H.F. Nijhout (1991) The Development and Evolution of Butterfly Wing Patterns. Smithsonian Institution Press, Washington and London.
7 ——(1856) A Letter to Wallace from the Amazon In A.C. Brackman (1980) A Delicate Arrangement: the strange case of Charles Darwin and Alfred Russel Wallace. Time-Life Books, New York.
8 Bowler, J. & Taylor, J. (1993) The Avifauna of Seram In I.D. Edwards, A.A. MacDonald & J. Proctor (1995)

Natural history of Seram, Maluku, Indonesia. Raleigh International, Intercept Ltd, Andover.

9   Brackman, A.C. (1980) A Delicate Arrangement: the strange case of Charles Darwin and Alfred Russel Wallace. Time-Life Books, New York.
10 Browne, J. (1983) The Secular Ark: Studies in the History of Biogeography. Yale University Press, New Haven.
11 Burrett, C., Duhig, N., Berry, R. & Varne, R. (1991) Asian and South-western Pacific Continental Terranes Derived from Gondwana, and their Biogeographic Significance In P.Y. Ladiges, C.J. Humphries & L.W. Martinelli (eds) (1991) Austral Biogeography. CSIRO, Australia.
12 Cheers, G. (1992) A Guide to Carnivorous Plants of the World. Angus and Robertson, Sydney.
13 Cox, C.B. & Moore, P.D. (1993) Biogeography: An Ecological and Evolutionary Approach. Blackwell Scientific Publications, Oxford.
14 Darwin, F. (ed.) (1888) Life and Letters of Charles Darwin In E. Mayr (1976) Evolution and the Diversity of Life: selected essays. Belknap Press, Cambridge.
15 Decker, R. & Decker, B. (1981) Volcanoes. Freeman & Company, New York.
16 Delson, E. (1980) Fossil macaques, phyletic relationships and a scenario of deployment In D.G. Lindburg (ed.) (1980) The macaques: studies in ecology, behaviour, and evolution. Van Norstrand Reinhold, New York.
17 Diamond, J.M. (1982) Mimicry of Friarbirds by Orioles. Auk 99: 187-196.
18 Duffels, J.P., & de Boer, A.J. (1990) Areas of endemism and composite areas in East Malesia In P. Baas, K. Kalkman & R. Geesink (eds) (1990) The Plant Diversity of Malesia. Kluwer Academic Publishers, Amsterdam.
19 Flannery, T.F. (1995a) Mammals of the South-West Pacific and Moluccan Islands. Reed Books, Sydney.

20 ——(1995b) The Future Eaters. Reed Books, Sydney.
21 George, W. (1964) Biologist philosopher: a study of the life and writings of Alfred Russel Wallace. Abelard-Schuman, London.
22 Gleick, J. (1987) Chaos. Cardinal, London.
23 Groombridge, B. (ed.) (1992) Global Biodiversity: Status of the Earth's Living Resources. Chapman and Hall, London.
24 Groves, C.P. (1981) Ancestors for the pigs: taxonomy and phylogeny of the genus *Sus*. Technical Bulletin No 3. Department of Prehistory, Research School of Pacific Studies. Australian National University, Canberra.
25 Hakluyt, R. (1903–5) The Principal Navigations Voyages Traffiques and Discoveries of the English Nation In George, W. (1964) Biologist philosopher: a study of the life and writings of Alfred Russel Wallace. Abelard-Schuman, London.
26 Heaney, L.R. (1985) Zoogeographic evidence for middle and late Pleistocene land bridges to the Philippine Islands. Modern Quaternary Research in South-East Asia, 9: 127–165.
27 ——(1986) Biogeography of mammals in South-East Asia: estimates of rates of colonisation, extinction and speciation. Biological Journal of the Linnean Society 28: 127–165.
28 Hooijer, D.A. (1975) Quaternary mammals west and east of Wallace's Line. Neth. J. Zool. 25: 46–56.
29 Horridge, A. (1986a) A Summary of Indonesian Canoe and Prahu Ceremonies. Second International Conference on Indian Ocean Studies, Perth, 5–12 December, 1984.
30 ——(1986b) The Evolution of Pacific Canoe Rigs. Journal of Pacific History, 21: 83–99.
31 Huxley, T.H. (1868) On the classification and distribution of the Alectoromorphae and Heteromorphae. Proceedings of the Zoological Soc. of London, 294–319.

32 Johnson, D.L. (1980) Problems in the land vertebrate zoogeography of certain islands and the swimming powers of elephants In L.R. Heaney (1985) Zoogeographic evidence for middle and late Pleistocene land bridges to the Philippine Islands. Modern quaternary research in South East Asia, 9: 127–165.
33 Jones, D.N., Dekker, R.W.R.J. & Roselaar, C.S. (1995) The Megapodes: Megapodiidae. Oxford University Press, Oxford.
34 Kaplan, G. & Roger, L. (1994) Orang-utans in Borneo. University of New England Press, Armidale.
35 Kemp, A.C. (1995) The Hornbills. Oxford University Press, Oxford.
36 Kirsch, J.A.W., Flannery, T.F., Springer, M.S. & Lapointe, F-J. (1995) Phylogeny of the Pteropodidae (Mammalia: Chiroptera) Based on DNA Hybridisation, with Evidence of Bat Morphology. Aust J. Zool. 43: 395–428.
37 Kitchener, D.J., Boeadi, Charlton, L. & Maharadatunkamsi (1990) Wild Mammals of Lombok. Records of the Western Australian Museum, Supplement No 33.
38 Knight W.J. & Holloway J.D. (1990) Insects and the Rain Forests of South East Asia (Wallacea). The Royal Entomological Society of London, London.
39 Kottelat, M., Whitten, A.J., Kartikasari, S.N. & Wirjoatmodjo, S. (1993). Freshwater Fishes of Western Indonesia and Sulawesi. Periplus, Singapore.
40 Lamarck, J.B. de (1914) Zoological philosophy: an exposition with regard to the natural history of animals (trans.) Hugh Elliot. Macmillan, London In E. Mayr (1976) Evolution and the Diversity of Life: selected essays. Belknap Press, Cambridge.
41 Lincoln, G.A. (1975) Bird counts either side of Wallace's Line. J. Zool. Lond. 177: 349–361.

42 MacKinnon, J. & Phillipps, K. (1993) A Field Guide to the Birds of Borneo, Sumatra, Java and Bali. Oxford University Press, Oxford.

43 Marchant, J. (1916) Alfred Russel Wallace: letters and reminiscences (2 Vols). Cassell, London.

44 Mayr, E. (1976). Evolution and the Diversity of Life: selected essays. Belknap Press, Cambridge.

45 Metcalfe, I. (1990) Allochthonous terrane processes in Southeast Asia. Phil. Trans. R. Soc. Lond. 331: 625–640.

46 ——(1993) Southeast Asian terranes: Gondwanaland origins and evolution In Findlay, Unrug, Banks & Veevers (eds) Gondwana. Balkema, Rotterdam.

47 ——(1994a) Gondwanaland origins, dispersion, and accretion of East and Southeast Asian continental terranes. Journal of South American Earth Sciences 7, 3/4: 333–347.

48 ——(1994b) Late Palaeozoic and Mesozoic Palaeogeography of Eastern Pangea and Tethys In Pangea: Global Environments and Resources. Canadian Society of Petroleum Geologists, Memoir 17: 97–111.

49 Miller, R. (1983). Continents in Collision. Time-Life Books, Amsterdam In Flannery, T. (1995) The Future Eaters. Reed Books, Sydney.

50 Moores, A. & Cotgreave, C. (1994) Sibley and Ahlquist's tapestry dusted off. Trends in Ecology and Evolution, 9:12, 455–500.

51 Musser G.G. (1987) The Mammals of Sulawesi In Whitmore, T.C. (ed.). Biogeographical Evolution of the Malay Archipelago. Clarendon Press, Oxford.

52 Myers L.C. (1978) Geomorphology, Chapter 4 In M. Luping, C. Wen & E.R. Dingley (eds) Kinabalu: Summit of Borneo: 91–94. Sabah Society Monograph 1978 cited in J.H. Beaman & R.S. Beaman (1989) Diversity and distrib-

ution patterns in the flora of Mount Kinabalu In P. Baas, K. Kalkman & R. Geesink (1989) The Plant Diversity of Malesia. Kluwer Academic Publishers, Amsterdam.

53 Nijhout, H.F. (1991) The Development and Evolution of Butterfly Wing Patterns. Smithsonian Institution Press, Washington and London.

54 Nowak, R.M. (1991) Walker's Mammals of the World. The John Hopkins University Press, Baltimore and London.

55 Raven, H.C. (1935) Wallace's Line and the distribution of Indo-Australian mammals. Bull. Amer. Mus. Nat. Hist. 68: 179–293.

56 Sclater, P.L. (1858) On the general geographical distribution of the members of the class Aves. J. Linn. Soc. (Zool.) Lond. 2: 130–145.

57 Sigurdsson, H. & Carey, S. (1989) Plinian and co-ignimbrite tephra fall from the 1815 eruption of Tambora volcano. Bull. Volcanol. 51: 243–270.

58 Simkin, T. & Fiske, R.S. (1983) Krakatau 1883: Eruption and its effects. Smithsonian Institution Press, Washington.

59 Simpson, G.G. (1977) Too many lines: the limit of the Oriental and Australian zoogeographic regions. Proceedings of the American Philosophical Society, 121: 107–20.

60 Slack, A. (1979) Carnivorous Plants. Doubleday, New York.

61 Smith, A.G., Hurley, A.M. & Briden, J.C. (1981) Phanerozoic paleo-continental world maps. Cambridge University Press, Cambridge.

62 Smith, J.M.B. (1977) Origins and ecology of the tropicalpine flora of Mount Wilhelmina, New Guinea. Biol. J. Linn. Soc. 9: 87–131.

63 Smith J.M.B. (1986) Origins and History of the Malesian High Mountain Flora In Vuilleumier, F. & Monasterio, M.

(eds) High Altitude Tropical Biogeography. Oxford University Press, Oxford.

64 Sondaar, P.Y. (1984) Fauna evolution and mammalian biostratigraphical of Java. Cour. Forsch. Inst. Senckenberg, 69: 219–35.

65 Sondaar, P.Y., van den Bergh, G.D., Mubroto, B. Aziz, F., de Vos, J. & Batu, U.L. (1994) Middle Pleistocene faunal turnover and colonization of Flores (Indonesia) by *Homo erectus*. Compte Rendus de l'Academie Des Science. t.319 Series II: 1255–1262.

66 van Bemmelen, R.W. (1949) The Geology of Indonesia. Government Printing Office, The Hague.

67 Vane-Wright, R.I. & Peggie, D. (1994) The Butterflies of Northern and Central Maluku: Diversity, Endemism, Biogeography, and Conservation Priorities. Tropical Biodiversity 2(1): 212.

68 Veevers, J.J. (1991) Phanerozoic Australia in the changing configuration of proto-Pangea through Gondwanaland and Pangea to the present dispersed continents In P.Y. Ladiges, C.J. Humphries & L.W. Martinelli (1991) Austral Biogeography. CSIRO. Australia.

69 Vickers-Rich, P. & Rich, T. (1993) Wildlife of Gondwana. Reed Books, Sydney.

70 von Koenigswald, G.H.R. (1967) An upper Eocene mammal of the family Anthracotheriidae from the island of Timor, Indonesia. Koninkl. Nederl. Akad. Wetensch. Proc. B-70: 528–33.

71 Wallace, A.R. (1859) Letter from Mr Wallace concerning the geographical distribution of birds. Ibis 1: 449–54.

72 ——(1869) The Malay Archipelago: The land of the Orang-utan and the bird of paradise. Macmillan, London.

73 ——(1876) The Geographical Distribution of Animals, Vol. 1. Macmillan, London.

74 ——(1881) Island Life [rev. edn 1911]. Macmillan, London.
75 Watts, C.H.S. & Baverstock P.R. (1994) Evolution in some South-east Asian Murinae (Rodentia), as assessed by microcomplement fixation of albumin, and their relationship to Australian murines. Aust. J. Zool. 42: 6, 711–22.
76 White, C.M.N. & Bruce, M.D. (1986) The Birds of Wallacea (Sulawesi, The Moluccas & Lesser Sunda Islands, Indonesia). British Ornothologists' Union, London.
77 Whitmore, T.C. (ed.) (1987) Biogeographical Evolution of the Malay Archipelago. Clarendon Press, Oxford.
78 Whitten, A.J., Damanik, S.J., Anwar, J. & Hisyam, N. (1987) The Ecology of Sumatra. Gadjah Mada University Press, Yogyakarta, Indonesia.
79 ——Mustafa, M. & Henderson, G.S. (1987) The Ecology of Sulawesi. Gadjah Mada University Press, Yogyakarta, Indonesia.
80 Williams, M.A.J. (1991) Evolution of the landscape In C.D. Haynes, M.G. Ridpath & M.A.J. Williams (eds) Monsoonal Australia: Landscape, ecology and man in the northern lowlands. Balkema, Rotterdam.
81 Wyllie, P.J. (1976). The Way the Earth Works: An Introduction to the New Global Geology and its Revolutionary Development. John Wiley & Sons, Inc., New York.

INDEX

maps in bold
figures italics

Aceros plicatus 201
Aepypodius 25
Agassiz, Louis 10, 11
Ahlquist, Jon 115
Ailurops ursinus 143
Alectura 25, 27
Allen, Charles 179
Alor islands 92
Amaurornis phoenicurus 104
Ambon 109, 217, 218
Anaphalis javanica 130
animals, geographic distribution 4, 12, 23, 40, 99
anoas 142, 149
Antarctica 85, 86
apes 139, 184-5, 194
Aru Islands 84, 86, 207, 210, 214, 215
Asian Golden Cat 184
Asian Golden Weaver 23
Austral warblers 114
Australasian Grebe 114
Australian Hobby 114
Australian Kestrel 114
Australian pesserines 114
Australian songbirds 117

babblers 114
Babirusa 138, 140-1, 147, 149, 154, 155
Babyrousa babyrussa 141, 154
Bacan 112, 115, 120, 145, 210
Bali 22-3, 31, 33, 36, 38, 68, 82, 86, 92, 99, 100, **101**, 102, 103-4, 108, 184
Banda Islands 69-71, 109, 210
Banda Sea 69, 70
Banda volcano 70
Banded Leafmonkey 80
bandicoot 120, 215, 216
Banteng Cow 92
barbets 24, 33, 100
Barking Deer 39, 117, 182
Bastin, S. Leonard 125
Bates, Henry Walter 2, 14, 34, 40, 159, 160

bats 105, 107, 144-5
Bear Cuscus 143, 147, 149, 154, 155
Bearded Pig 152
beetles 181, 210
Belideus ariel 120
bird of paradise 3, 113, 114, 116, 212-16
birds
   Australian 38, 100-1, 102-3, 113-14, 116, 121
   distribution 32-3, 100-5, 106, 107, 115-16
   DNA hybridisation 115-16
   family tree 115
   fossils 116
   migration 107-8, 114-15, 116-17
   Oriental 38, 99-101, 104, 107, 108, 113, 123
   songbirds 114-15, 116, 117, 118
   visual mimicry 110-11
black grasshopper-thrush 102
Black Rat 119
blue pitta 103
Bornean Bay Cat 184
Borneo 31, 33, 58, 62, 63, 84, 86, 92, 94, 100, 102, 121, 133, 135, 136, 137, **144**, 155, 156, 157, 167, 178, 182, 184, 190, 191-2, 193, 195, 196, 210, 215
Borneo Bristlehead 116
Brown Prinia 102
brush-turkeys 25, 27, 32
Bubalus depressicornis 142
Bubalus quarlesi 142
Buceros bicornis 198-201
buffalo 145
Buru 110, 112, 113
butcherbirds 116
butterflies 14, 63, 106, 108, 112, 113, 157, 160-75, *162, 165, 168, 172*, 181, 210, 220
Button-quails 114

cassowary 114
Ceyx cajeli 110
Charmosyna placentis 109
chestnut 129

229

chimpanzee 184, 194, 195
Christian concept of creation 4
cicadas 64-7, 155
Cicinnurus reginus 212
civets 33, 39, 137, 144, 145, 152, 184
climate change 64, 85-6, 89-91
Clouded Leopard 184, 189
cockatoos 24, 32, 105
Collard Kingfisher 114
comb-crested Jacana 104
Common Chimpanzee 184
Common Malay Civet 144
Common Palm Civet 117
Common Spotted Cuscus 215
Common Tailorbird 102
Continental Drift 43-4, 45-65, **52, 53, 54, 55, 56, 57, 59, 60, 61, 62,** 138-9, 149, 150
cranberry 129
Crocidura fulginosa 148
Crocidura levicula 148
Crocidura monticola 148
Crocidura nigriceps 148
Crocidura rhoditis 148
crow 114, 116
cuckoo-shrikes 116
Cuon alpinus 39
currawongs 116
cuscus 33, 117, 120, 143, 181, 215
Cuscus ornatus 120
Cynocephalus variegatus 181

da Gama, Vasco 70
daisies 128, 130
Darwin, Charles 11, 12, 13, 15-18, 41, 76, 83, 124, 136, 159, 207
  theory of evolution 18
Darwin, Erasmus 8, 9
de Houtman, Cornelis 71
deer 39
dhole 184
Diamond, Jared 111
Dingo 39
Ducula bicolor 109
durian tree 188-9

earthquakes 72-3
echidna 32
elephant 65, 91, 92, 139, 192, 198
Elephas celebensis 93
emu 114
Eumyias panayensis 104
Eurystomus azureus 110

evolution 9-10, 13-15, 16, 18-19, 23, 160, 207, 220

fantails 114
fauna
  Australian 32, 104
  boundaries 34
figbird 114
fish 61-2, 191, 192
Fishing Cat 184
Flat-headed Cat 184
flora 32, 93-5, 128-34
Flores 36, 86, 92, 93, 106-7, 118, 210
flycatcher 100, 105, 110, 114
Flying Lemur 181, 182
Footless Paradise Bird 214
Forbes, Edward 13
Forest Kingfisher 114
fossils 91, 93, 116, 120, 141, 142
foxglove 129
friarbirds 100, 110-12
fruit-bats 118
fruit-dove 109, 114
fruit-pigeon 109
fruit-thrushes 24, 33

Galapagos Islands 15-16, 84, 105
Gaultheria 128, 132
Gebe Cuscus 120
Giant Pangolin 92
gibbons 40, 183
giraffe 39
glacial cycles 85, 86, 87-91
Gleichenia arachnoidea 130
Gondwana 31, **47**, 47, 50-9, **52, 53, 54, 55, 56, 59,** 116, 124, 137, 154
Gorilla 184, 194
Gorilla gorilla 184
Graphium androcles 220
Gray, Asa 17
Great Pied Hornbill 198-201
Greater Bird of Paradise 213, 214-16
Green Peacock 102
ground starling 102

Halmahera 109, 110, 120, 150
Haloragis micrantha 131
Harpactes reinwardtii 100
hedgehogs 39
Helmeted Friarbird 24
hippopotamus 39, 91, 137, 141
Homo sapiens 184, 194-5

honeyeater 24, 32, 33, 38, 110, 111, 112, 114, 117
honeysuckle 107, 129, 130
Hooker, Joseph 12
Huxley, Thomas Henry 35, 37
Huxley's Line 37, 157
hyaena 91, 139
Hyosciurus heinrichi 147

Indonesia 70, 76, 94, 121
Irediparra gallinacea 104
Island Verditer Flycatcher 104

Jacana 114
jackdaws 116
Java 31, 58, 63, 84, 92, 94, 99, 100, 102, 104, 121, 128, 136, 144, 156, 182, 184, 192, 215
Javan Rhinoceros 92
jays 116, 121

Kai island 203, 207, 209, 210, 219
Kallima inachus 172
kangaroos 32, 215
Kaplan, Gisela 187, 189
King Bird of Paradise 212, 213
kingfisher 105, 109, 110, 114, 117
Kitchener, Danny 118
Kookaburra 114
Krakatau eruption 76-79, 80-82

La Peyrere, Issac de 5, 7
Lamark, Jean-Baptiste de 8, 9-10
leaf-thrushes 33
Leipoa 25
Lemur 13, 39, 43, 137, 138, 139, 182, 192
Leopard Cat 117, 184
leopards 117, 184
Lesser Birds of Paradise 216
Lesser Sundas 56, 68, 99, 157
Lincoln, G. A. 38
Linnaeus, Carolus 5-7, 212, 213
Linnean Society 18
Little Civet 117
Lombok 22, 23, 32, 33, 36, 38, 68, 83, 86, 92, 95, 100, **101**, 102, 103-4, 106, 108, 117, 118
Long-tailed Macaque 117, 152
Lowland Anoa 142
Lycocorax pyrrhopterus 110
Lydekker's Line 35, 37
Lyell, Sir Charles 14-15, 17, 32, 42-3

Macac maura 147
Macaca nemestrina 146
Macac nigra 147
Macaca nigrescens 145
Macaca tonkeana 147
macaques 39, 117, 145-7, 183
Macadamia hildebrandii 129
Macrogalidia musschenbroekii 144
Madagascar 137, 140, 154, 155
magnolias 129
magpies 116
Makassar 22
Makassar Strait 33, 34, 63, 66, 92, 95
Malay Archipelago 2, 18, 34, 69, 91, 94, 138, 197, 201, 210
Malay Bear 92
Malay Peninsula 86, 99, 106, 124, 157, 192
Malay Tapir 92, 182
Malayan Sun Bear 184
Maleo Macrocephalon maleo 25, 28-9
malleefowl 25
Malthus, Thomas Robert 16
Maluka 30, 100, 104, 105, 106, 108, 112, 117, 120, 157
mammals 93, 117, 144-5, 152-3
Mammuthus primigenus 90
Manis javanica 182
maples 129
Marbled Cat 184
Mariana Islands 31
Mastodon 90
Mayr, Ernst 37
megapode family 25-31, 26, 114
Megapode Eulipoa 25
Megapodius cumingii 27, 31
Megapodius eremita 28
Megapodius laperouse 31
Megapodius nicobariensis 31
Megapodius reinwardt 24, 25
Megapodius wallacei 30
Mendel, Gregor 169
Metallic Starling 114
Metroxylon sagu 205
Mindanao 150-1, 153, 156
Monarcha loricata 110
monarchs 114
monkeys 33, 40, 139, 183, 192, 194
Morotai 105, 110, 113
Mount Gebe 129
Mount Kerinci 130
Mount Kinabalu 133
Mount Pangrango 129, 132
Mount Wilhelm 132

Mountain Anoa 142
Mountain Tailorbird 104
Muller, Salomon 35-6
Muller's Line 34-5
Murray's Line 35, 36

Nasalis larvatus 40, 190
natural selection 16, 18, 160, 170, 175, 207
Nectarinea auriceps 110
Nectarinea proserpina 110
Nepenthes pitcher plants 124-28
New Guinea 53, 54, 56, 59, 66, 67, 89, 94, 108, 131, 133, 215
Nicobar Islands 30, 31
Northern Common Cuscus 215
Nusa Tenggara 56, 58, 63, 67, 68, 81, 86, 92, 93, 95, 99, 106, 107, 117, 118-19, 157
nutcrackers 116

oak 129
Obi Cuscus 120
Olive-backed Oriole 114
opossum 122, 143
Orang-utan 3, 13, 40, 80, 91, 92, 177-96, 197, 198, 220
Orange-footed Megapode 25, 27, 31
Oriental Bulbul 38
Oriental civet 152
Oriental tree shrews 181
oriole 100, 102, 110-12, 114, 116
Oriolus bouroensis 110
Oriolus cruentus 100
Ornithoptera poseidon 163
Ornithoptera croesus 164
Orthotomus cuculatus 104
otters 33
owls 114

palm civet 152, 184
Pan paniscus 184
Pan troglodytes 184
Pangaea 43, 47, 50-1
pangolins 39, 117, 139, 152, 181, 182
panther 92
Papilio aenomaus 169
Papilio androcles 220
Papilio liris 168
Papilio memnon 173, 174, 175
Papuan Wreathed Hornbill 201
Paradisaea apoda 213
Paradisaea minor 216

Paradise Crow 110
paradise-kingfisher 114
parrots 33, 36, 100, 110, 114, 116, 117
Peelerophon oehlerti 51
Pericrocotus miniatus 100
Phalanger orientalis 120
Phalanger rothschildi 120
pheasants 32, 121
Philemon buceroides 24
Philemon fuscicapillus 110
Philemon moluccensis 110
Philippines 137, 149, 152, 154, 156-7, 193
pigeons 105, 116, 117
pigs 91, 141, 142, 145
Pitta elegans 103
plains-wanderer 114
Plantain Squirrel 117
platypus 32
Pleistoncene glaciations 85-95, **87**, 100, **101**, 107, 119, 133, 134, 146, 149, 150, 153, 156
Ploceus hypoxanthus 23
Pongo pygmaeus 184
porcupines 139, 152
possums 32, 215
Primula imperialis 130-132
Primula prolifera 131
Proboscis Monkey 40, 190
Project Wallace 211
Prosciurillus abstrusus 147
Prosciurillus leucomus 147
Prosciurillus murinus 147
Ptilonopus regina 109
Pygmy Chimpanzee 184
python 189

racquet-tailed kingfisher 109, 110
Rafflesia 3
rats 119
Rattus norvegicus 119
Raven, R. C. 38-9, 63
ravens 116
Red-backed Kingfisher 114
Rensch, B. 38
rhinoceros 39, 91, 92, 139, 192
Rhododendron 128, 129, 132
Rhynchomeles prattorus 120
robins 114
Rogers, Lesley 187, 189
rooks 116
rosy barbet 102
Rubrisciurus rubriventer 147

Rufous Night Herron 114

Sacred Kingfisher 114
Sago tree 205-6
Sarawak 190, 192
Sarawak Law 13, 14, 17, 23
Schwanwitsch, B. N. 161
Sclater, W. L. and P. L. 37, 99, 106
Sclater's Line 35, 37
Semioptera wallacii 113
Senecio sumatrana 130
Seram 59, 110, 112-13, 119, 215, 218
shrew 39, 145, 149, 181-2, 192
Sibley, Charles 115
Silvered Leaf Monkey 117
Simpson, George Gaylord 35, 38
Slow Loris 40, 183
Small Sulawesi Cuscus 143, 149
species, origin 12, 14, 15-19, 160-61
Spilocuscus macultaus 120
Spotted Harrier 114
squirrels 33, 39, 145, 147, 149, 150, 183, 192
Sri Lanka 84
St John's-wort 130
Standard-wing Bird of Paradise 113
Stegadon trigonocephalus 93
stegodons 91, 92, 93
Strigocuscus celebensis 143
Strigocuscus pelengensis 144
Süffert, F. 161
Sulawesi 29, 31, 33, 36, 56, 58, 59, 62, 63, 66-7, 86, 92, 93, 105, 108, 119, 120, 128, 135-7, 161, 211, 215, 219, 220
Sulawesi Palm Civet 144
Sulawesi-Sangihe volcanic arc 70
Sulawesian Long-nosed Ground Squirrel 147
Sulu islands 150
Sumatra 31, 63, 84, 86, 92, 94, 99, 100, 121, 144, 156, 156-7, 182, 191, 193, 197, 198, 215
Sumatran Rabbit 80
Sumatran Rhinoceros 92
Sumatran Tiger 184
Sumbawa 36, 38, 106
Sun Bear 39
sunbirds 117
Sunda arc 69, 72
Sunda Island Porcupine 117
Sunda Shelf islands 144, 147, 148-9, 157, 182, 183, 184, 190, 193

Sunda Shrew 117
Sus barbatus 152
Sus celebensis 142, 149
symmetry systems 161, 170

Talegalla 25
Tambora eruption 80
Tansysiptera acis 110
Tansysiptera doris 110
Tansysiptera galatea 109
Tansysiptera isis 110
tapir 39, 80, 91, 192
tarsier 80, 138, 139-40, 142, 145, 147, 149, 154, 183
Tarsius bancanus 139
Tarsius pumilis 139
Tarsius spectrum 139
Tarsius syrichta 139
tectonic plate
  Asian 149, 150
  Australian 69-70, 73, 149, 150
  cross section 75
  Eurasian 70, 73
  Indo-Australian 69-70, 73
  map **46**
  movement 72-3, 75, 149
  theory 45-68
Ternate 97-98
Thomas' Leafmonkey 80
three-toed woodpecker 102
thrushes 114, 121
tigers 91, 92, 183-4
Timor 58, 86, 92, 93, 100, 107, 108, 117
Toba eruption 80
treecreepers 114
trillers 116
Trogonoptera brookiana 167, 168
trogons 33, 100

Vaccinium 128, 132
van Bemmelen, Reinout W. 45
violets 128, 129
Viverra tangalunga 144, 152
Vogelkop Peninsula 59, 66, 67, 84
volcanoes **46**, 67-8, 70, 72, 73-75, 79-80, 81
von Humbolt, Alexander 7

Wallace, Alfred Russel
  bird observations 113-14
  in Borneo 177-81
  on changing sea levels 84
  character 1-2, 3, 11, 180

'Eastern Collections' 181, 207
education 11
on evolution 10-11, 15, 18, 23,
　160, 207
on geographic distribution of
　animals 23, 39-40
in Malay Archipelago 2, 3,
　11-12, 18, 34, 71, 197, 201,
　206, 210
in Seram 112-13
in Ternate 97-99
theory of evolution 18, 23, 220
visits Amazon 2, 3, 12
Wallace Line 23, 33-7, 38-40, 62, 63,
　66, 68, 82, 84, 95, 100, 103, 156,
　157, 192, 207, 220
　map **26, 35**

Weber's Line 35, 37
Wegener, Alfred 43-5
white cockatoos 102
White-bellied Cuckoo-shrike 114
White-breasted Waterhen 104
white-eyes 117
Whitehanded Gibbon 80
wild-cats 33
wombats 32
wood-peckers 24, 32, 33, 121
wood-sorrel 129
wood-swallows 114, 116
Woolly Mammoth 90
wormwood 129

yellow-headed weaver 102